Table of Contents

100546 61500

GETTING STARTED WITH STATA
FOR WINDOWS®
RELEASE 10

A Stata Press Publication
StataCorp LP
College Station, Texas

Stata Press, 4905 Lakeway Drive, College Station, Texas 77845

The suggested citation for this software is

StataCorp. 2007. *Stata Statistical Software: Release 10*. College Station, TX: StataCorp.

Cross-Referencing the Documentation

When reading this manual, you will find references to other Stata manuals. For example,

[U] **26 Overview of Stata estimation commands**
[R] **regress**
[D] **reshape**

The first is a reference to chapter 26, *Overview of Stata estimation commands* in the *Stata User's Guide*, the second is a reference to the `regress` entry in the *Base Reference Manual*, and the third is a reference to the `reshape` entry in the *Data Management Reference Manual*.

All the manuals in the Stata Documentation have a shorthand notation:

[GSM]	*Getting Started with Stata for Macintosh*
[GSU]	*Getting Started with Stata for Unix*
[GSW]	*Getting Started with Stata for Windows*
[U]	*Stata User's Guide*
[R]	*Stata Base Reference Manual*
[D]	*Stata Data Management Reference Manual*
[G]	*Stata Graphics Reference Manual*
[P]	*Stata Programming Reference Manual*
[XT]	*Stata Longitudinal/Panel-Data Reference Manual*
[MV]	*Stata Multivariate Statistics Reference Manual*
[SVY]	*Stata Survey Data Reference Manual*
[ST]	*Stata Survival Analysis and Epidemiological Tables Reference Manual*
[TS]	*Stata Time-Series Reference Manual*
[I]	*Stata Quick Reference and Index*
[M]	*Mata Reference Manual*

Detailed information about each of these manuals may be found online at

http://www.stata-press.com/manuals/

About this manual

This manual discusses **Stata for Windows**®. Stata for Macintosh® users should see *Getting Started with Stata for Macintosh*; Stata for Unix® users should see *Getting Started with Stata for Unix*. This manual is intended both for people completely new to Stata and for experienced Stata users new to Stata for Windows. Previous Stata users will also find it helpful as a tutorial on some new features in Stata for Windows.

Following the numbered chapters are four appendices with information specific to Stata for Windows.

We provide several types of technical support to registered Stata users. Chapter 6 of this manual describes the resources available to help you learn about Stata's commands and features. One of these resources is the Stata web site (http://www.stata.com), where you will find answers to frequently asked questions (FAQs), as well as much other useful information. If you still have questions after looking at the Stata web site and the other resources described in chapter 6, you can contact us as described in [U] **3.9 Technical support**.

Using this manual

The new user will get the most out of this book by treating this book as an exercise book, working through each example at the computer. The material builds, so that material from earlier chapters will often be used in later chapters. Bear in mind that Stata is a rich and deep statistical package—just like statistics itself. The time spent working the examples will be repaid with dividends when doing true statistical analyses.

The experienced user may still have something to learn from this manual, despite its name. Looking through the chapters to see if there is anything new or forgotten can help.

1 Installation

Stata for Windows

Description Stata for Windows is available for Windows 2000, XP, and Vista (32-bit x86 and 64-bit x86-64 and Itanium®).

(Versions for Macintosh and for Unix are also available.)

Stata/MP Professional version of Stata.
Parallel-processing-capable Stata/SE.
Fastest on multiple-core/multiple-processor machines.
Maximum of 32,766 variables; observations limited only by computer memory.
String variables up to 244 characters.
Matrices up to $11,000 \times 11,000$ in Stata. In Mata, limited only by memory.

Stata/SE Professional version of Stata.
Fastest on a single-processor machine.
Maximum of 32,766 variables; observations limited only by computer memory.
String variables up to 244 characters.
Matrices up to $11,000 \times 11,000$ in Stata. In Mata, limited only by memory.

Stata/IC Professional version of Stata.
Very fast.
Maximum of 2,047 variables; observations limited only by computer memory.
String variables up to 244 characters.
Matrices up to 800×800 in Stata. In Mata, limited only by memory.

Small Stata Stata for small computers.
Slower than Stata/MP, Stata/SE, and Stata/IC.
Maximum of 99 variables and approximately 1,000 observations.
String variables up to 200 characters.
Matrices up to 40×40.

(*Continued on next page*)

Upgrade or update?

If you use Stata 9 or an earlier release and you would like to upgrade to Stata 10, or if you have never installed Stata before on this computer, this is the chapter for you. If you have already installed Stata 10 and you would like to install the latest updates to Stata 10, see chapter 20.

Upgrading to Stata/MP, Stata/SE, or Stata/IC

If you have already installed a flavor of Stata 10 and have purchased an upgrade to Stata/MP, Stata/SE, or Stata/IC, run the installer again. Check **Modify**, click **Next**, and choose the flavor of Stata you wish to install. Your existing copy will not be affected.

Make sure that you have your *License and Authorization Key* before doing this. After installation, you will immediately want to update your executable because it will most likely be out of sync with your ado-files.

Before you install

Before you begin the installation procedure:

1. Make sure you have the Stata Installation CD.
2. Make sure you have a *License and Authorization Key*.
3. Determine from the *License and Authorization Key* whether you should install Stata/MP, Stata/SE, Stata/IC, or Small Stata. If you want to install a 64-bit version of Stata, your license must say that it is a 64-bit license and you must perform the installation procedure on a 64-bit version of Windows.
4. Decide where you want to install the Stata software. We recommend C:\Program Files\Stata10. If you want to install Stata on a network drive, see [GSW] **D.3 Installing Stata for Windows on a network drive**.
5. Decide where you want to set the working directory. This should be different from the installation directory so that files that you create will not get mixed up with Stata's files. We recommend C:\data.
6. If you already have an old version of Stata on your system, decide whether you want to keep it or uninstall it. We do not recommend having more than one version of Stata installed at a time.

Installation

Be sure that Windows is installed and properly running before you attempt to install Stata for Windows. Have your Stata *License and Authorization Key* with you.

1. Insert the CD in the CD-ROM drive.
2. If you have **Auto-insert Notification** enabled, the installer will start automatically. Otherwise, select **Run** from the Windows **Start** menu, and enter D:\setup.exe (assuming that D:\ is the drive letter for your CD-ROM) to start the installer.
3. You will see an opening screen of information. After reading this screen carefully, click **Next**.
4. The installer will display options for personalizing your installation and making Stata accessible to all users that share your computer. Make any necessary changes, and click **Next**.

5. At the **Select Components** dialog, you can choose which flavor of Stata to install. If you are running the installer on a 64-bit version of Windows and your license allows it, you can choose to install a 64-bit version of Stata/MP, Stata/SE, or Stata/IC. Your license must say that it is a 64-bit license if you want to install a 64-bit version of Stata.

6. The installer will ask you where you want to install Stata.

    a. We recommend that you choose the default directory—C:\Program Files\Stata10.

    b. If you want to install Stata on a network drive, see [GSW] **D.3 Installing Stata for Windows on a network drive** before continuing.

    c. When you have chosen an installation directory, click **Next**.

7. The installer will then ask you where you want to set the default working directory.

    a. The default working directory is the default location for your datasets, graphs, and other Stata-related files.

    b. We recommend that you choose the default directory—C:\data.

    c. When you have chosen a default working directory, click **Next** to begin the installation.

8. When the installation is complete, click **Finish** to exit the installer.

9. If you would like to modify your installation or install other flavors of Stata that are below the flavor of your license, you can run the installer a second time. It will bring up the Application Maintenance dialog. Check **Modify**, and click the **Next** button. You can then choose the flavor(s) of Stata that you would like to add.

(Continued on next page)

Initialize the license

Important note to Windows Vista users: the first time you run Stata, you **must** run the application as an administrator to make sure that future updates run smoothly. To to this, you must go to where you installed Stata (typically C:\Program Files\Stata10), right-click on the Stata application icon, and select **Run as administrator**. You need to do this when you first install Stata and any time you wish to update Stata (see chap. 20).

The first time that you start any version of Stata, it will prompt you for the information on your *License and Authorization Key*. You must enter something for all fields in the dialog before you can continue. The code and authorization are not case sensitive. If you make a mistake typing the codes, you will be prompted to try again.

If you get the message "The serial number, code, and authorization are inconsistent!", try the initialization again. Be careful when typing your code and authorization key. Anything that looks like "o" is the letter *oh*, anything like "0" is a zero, anything like "1" is the number one, and anything like "L" is the letter *el*.

Important: Do not lose your paper license. Keep it in a safe place. You may need it again in the future.

Update Stata if necessary

StataCorp releases updates to Stata often. These updates may include new features and bug fixes that can be automatically downloaded and installed by Stata from the Internet. There may be updates to Stata more recent than the version of Stata on your CD.

The first time Stata is launched, a dialog will open asking if you would like to check for updates now. Click **OK** to do so. If an update is available, follow the instructions. If you have trouble connecting to the Internet from Stata, visit http://www.stata.com/support/faqs/web/ for help. See chapter 2 for information on starting and exiting Stata.

If you are using Windows Vista and you run into trouble with the update, it might be because you are not running Stata as an administrator. The first time you run Stata, you **must** run the application as an administrator to make sure that future updates run smoothly. To to this, you must go to where you installed Stata (typically C:\Program Files\Stata10), right-click on the Stata application icon, and select **Run as administrator**.

It's also a good idea to periodically check for updates to Stata. Stata can automatically check for updates for you. See chapter 20 for more information about updating.

Register your copy

Make sure that your copy of Stata is registered. As a registered Stata user, you are entitled to free technical assistance should you have any questions, and we will keep you informed of any new products or advancements that have been announced. To register your copy of Stata, fill out the online registration form at http://www.stata.com/register/, or return the registration card that came with your Stata software.

Documentation

After installing, take the time to check that you have all your documentation.

Whether you purchased Stata/MP, Stata/SE, Stata/IC, or Small Stata, you should have received
Single-sheet *License and Authorization Key*
Installation CD
Registration card

A Base Stata Documentation Set consists of the following:
Getting Started with Stata for Windows (this book)
Stata User's Guide
Stata Base Reference Manual (3 volumes)
Stata Data Management Reference Manual
Stata Graphics Reference Manual
Stata Quick Reference and Index

A Complete Stata Documentation Set consists of the following:
Getting Started with Stata for Windows (this book)
Stata User's Guide
Stata Base Reference Manual (3 volumes)
Stata Data Management Reference Manual
Stata Graphics Reference Manual
Stata Longitudinal/Panel-Data Reference Manual
Stata Multivariate Statistics Reference Manual
Stata Programming Reference Manual
Stata Survey Data Reference Manual
Stata Survival Analysis and Epidemiological Tables Reference Manual
Stata Time-Series Reference Manual
Stata Quick Reference and Index
Mata Reference Manual (2 volumes)

Notes

2 Starting and exiting Stata

Starting Stata

To start Stata:

1. Select it from the Windows **Start** menu.

2. Or double-click on a Stata data file. Stata data files are data files created by Stata and have the extension .dta. When you double-click on a Stata data file, Stata is started and the data file is loaded into Stata; see chapter 7.

When Stata has launched, you will see

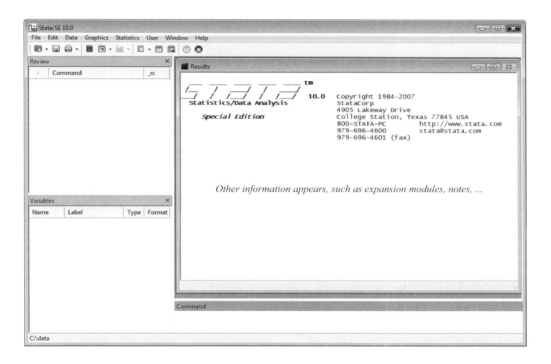

Note

There are other ways to start Stata. You may find one of these ways more convenient. Please read [GSW] **C. Advanced Stata usage** at the end of this manual for more information.

Exiting Stata

To exit Stata, select **Exit** from the **File** menu or press Alt-F4. If you have made any changes to the data in your dataset, including creating a new dataset, you will be prompted to save the changes.

If you exit Stata by typing the equivalent exit command, you will see the following in the Results window if you have made changes to your data:

```
. exit
no; data in memory would be lost
r(4);
```

If you would like to save your changes, you could save the dataset as you save any other file by using **File > Save**, or Ctrl-S. You can also save the dataset by typing save *filename*. You could then type exit again, and Stata will quit. If you do not wish to save your changes, you can force Stata to exit without saving the dataset by typing exit, clear.

If you experience any problems when trying to start Stata, see [GSW] **A. Troubleshooting Stata**.

3 Introducing Stata—sample session

Introducing Stata

Now that you have Stata installed and registered, you probably would like to use it. This chapter will run through a sample work session, introducing you to a few of the basic tasks that can be done in Stata, such as opening a dataset, investigating the contents of the dataset, using some descriptive statistics, making some graphs, and doing a simple regression analysis. As you would expect, we will only brush the surface of many of these topics. This approach should give you a sample of what Stata can do and how Stata works. There will be brief explanations along the way, with references to chapters later in this book, as well as to the online help and other Stata manuals. We will run through the session by using both menus and dialogs and Stata's commands, so that you can gain some familiarity with both.

So, take a seat at your computer, put on some good music, and work along with the book.

Sample session

The dataset that we will use for this session is a set of data about vintage 1978 automobiles sold in the United States.

To follow along using point-and-click, note that the menu items are given by **Menu > Menu Item > Submenu Item > etc**. To follow along using the Command window, type the commands that follow a dot (.) in the boxed listings below into the small window labeled **Command**. When there is something of note about the structure of a command, it will be pointed out as a **Syntax note**.

Start by loading the `auto` dataset, which is included with Stata. To use the menus,

1. Select **File > Example Datasets...**.
2. Click on **Example datasets installed with Stata**.
3. Click on **use** for `auto.dta`.

The result of this command is threefold:

- Some output appears in the large Results window.

```
. sysuse auto
(1978 Automobile Data)
```

The output consists of a command and its result. The command is in white and follows the period (.): `sysuse auto`. The output is in green here and is a brief description of the dataset. **Note:** if a command intrigues you, you can type `help` *commandname* in the Command window to find help. If you want to explore at any time, **Help > Search...** can be informative.

- The same command, `sysuse auto`, appears in the small Review window to the upper left. The Review window keeps track of commands Stata has run, successful and unsuccessful. The commands can then easily be rerun. See chapter 4 for more information.

- A series of variables appears in the small Variables window to the lower left.

9

You could have opened the dataset by typing `sysuse auto` followed by *Enter* into the Command window. Try this now. `sysuse` is a command that loads (uses) example (system) datasets. As you will see during this session, Stata commands are often simple enough that it is faster to use them directly. This will be especially true once you become familiar with the commands you use the most in your daily use of Stata.

Syntax note: in the above example `sysuse` is the Stata command, whereas `auto` is the name of a Stata data file.

Simple data management

We can get a quick glimpse at the data by looking at it in the **Data Browser**. This can be done by clicking the Data Browser button, 🔍 , or selecting **Data > Data Browser (read-only Editor)** from the menus, or typing the command `browse`.

When the Data Browser window opens, you can see that Stata regards the data as one rectangular table. This is true for all Stata datasets. The columns represent *variables*, whereas the rows represent *observations*. The variables have somewhat descriptive names, whereas the observations are numbered.

	make	price	mpg	rep78	headroom	trunk
1	AMC Concord	4,099	22	3	2.5	
2	AMC Pacer	4,749	17	3	3.0	
3	AMC Spirit	3,799	22	.	3.0	
4	Buick Century	4,816	20	3	4.5	
5	Buick Electra	7,827	15	4	4.0	
6	Buick LeSabre	5,788	18	3	4.0	
7	Buick Opel	4,453	26	.	3.0	
8	Buick Regal	5,189	20	3	2.0	
9	Buick Riviera	10,372	16	3	3.5	
10	Buick Skylark	4,082	19	3	3.5	
11	Cad. Deville	11,385	14	3	4.0	
12	Cad. Eldorado	14,500	14	2	3.5	
13	Cad. Seville	15,906	21	3	3.0	
14	Chev. Chevette	3,299	29	3	2.5	
15	Chev. Impala	5,705	16	4	4.0	

The data are displayed in multiple colors—at first glance it appears that the variables listed in black are numeric, whereas those that are in colors are text. This is worth investigating. Click on a cell under the `make` variable: the input box at the top displays the make of the car. Scroll to the right until you see the `foreign` variable. Click on one of its cells. Although the cell may display "Domestic", the input box displays a 0. This shows that Stata can store categorical data as numbers but display human-readable text. This is done by what Stata calls *value labels*. Finally, under the `rep78` variable, which looks to be numeric, there are some cells containing just a period (.). As we will see, these correspond to missing values.

Syntax note: here the command is `browse` and there are no other arguments.

Looking at the data in this fashion, though comfortable, lends little information about the dataset. It would be useful for us to get more details about what the data are and how the data are stored. Close the Data Browser by clicking its close button—while it is open, we cannot give any commands to Stata.

We can see the structure of the dataset by *describing* its contents. This can be done either by going to **Data > Describe data > Describe data in memory** in the menus and clicking **OK** or by typing `describe` in the Command window and pressing *Enter*. Regardless of which method you choose, you will get the same result:

```
. describe
Contains data from C:\Stata10\ado\base\a\auto.dta
  obs:            74                          1978 Automobile Data
  vars:           12                          13 Apr 2007 17:45
  size:        3,478 (99.9% of memory free)   (_dta has notes)

              storage   display    value
variable name   type    format     label      variable label

make            str18   %-18s                 Make and Model
price           int     %8.0gc                Price
mpg             int     %8.0g                 Mileage (mpg)
rep78           int     %8.0g                 Repair Record 1978
headroom        float   %6.1f                 Headroom (in.)
trunk           int     %8.0g                 Trunk space (cu. ft.)
weight          int     %8.0gc                Weight (lbs.)
length          int     %8.0g                 Length (in.)
turn            int     %8.0g                 Turn Circle (ft.)
displacement    int     %8.0g                 Displacement (cu. in.)
gear_ratio      float   %6.2f                 Gear Ratio
foreign         byte    %8.0g      origin     Car type

Sorted by:  foreign
```

If your listing stops short, and you see a blue —more— at the base of the Results window, pressing the space bar or clicking on the blue —more— itself will allow the command to be completed.

At the top of the listing, some information is given about the dataset, such as where it is stored on disk, how much memory it occupies, and when the data were last saved. The bold `1978 Automobile Data` is the short description that appeared when the dataset was opened and is referred to as a *data label* by Stata. The phrase `_dta has notes` informs us that there are notes attached to the dataset. We can see what notes there are by typing `notes` in the Command window:

```
. notes
_dta:
  1.  from Consumer Reports with permission
```

Here we see a short note about the source of the data.

Looking back at the listing from `describe`, we can see that Stata keeps track of more than just the raw data. Each variable has the following:

1. A *variable name*, which is what you call the variable when communicating with Stata.

2. A *storage type*, which is the way in which Stata stores its data. For our purposes, it is enough to know that types beginning with `str` are *string*, or text, variables, whereas all others are numeric. See [U] **12 Data**.

3. A *display format*, which controls how Stata displays the data in tables. See [U] **12.5 Formats: controlling how data are displayed**.

4. A *value label* (possibly). This is the mechanism that allows Stata to store numerical data while displaying text. See chapter 10 and [U] **12.6.3 Value labels**.

5. A *variable label*, which is what you call the variable when communicating with other people. Stata uses the variable label when making tables, as we will see.

A dataset is far more than simply the data it contains. It is also information that makes the data usable by someone other than the original creator.

Although describing the data says something about the structure of the data, it says little about the data themselves. The data can be summarized by clicking **Statistics > Summaries, tables, and tests > Summary and descriptive statistics > Summary statistics**, and clicking the **OK** button. You could also type summarize in the Command window and press *Enter*. The result is a table containing summary statistics about all the variables in the dataset:

```
. summarize
    Variable |       Obs        Mean    Std. Dev.       Min        Max

        make |         0
       price |        74    6165.257    2949.496       3291      15906
         mpg |        74     21.2973    5.785503         12         41
       rep78 |        69    3.405797    .9899323          1          5
    headroom |        74    2.993243    .8459948        1.5          5

       trunk |        74    13.75676    4.277404          5         23
      weight |        74    3019.459    777.1936       1760       4840
      length |        74    187.9324    22.26634        142        233
        turn |        74    39.64865    4.399354         31         51
displacement |        74    197.2973    91.83722         79        425

   gear_ratio |        74    3.014865    .4562871       2.19       3.89
     foreign |        74    .2972973    .4601885          0          1
```

From this simple summary, we can learn a bit about the data. First of all, the prices are nothing like today's car prices—of course, these cars are now antiques. We can see that the gas mileages are not particularly good. Automobile aficionados can gain some feel for other characteristics.

There are two other important items here:

The variable make is listed as having no observations. It really has no numerical observations, because it is a string (text) variable.

The variable rep78 has 5 fewer observations than the other numerical variables. This implies that rep78 has five missing values.

Although we could use the summarize and describe commands to get a bird's eye view of the dataset, Stata has a command that gives a good in-depth description of the structure, contents, and values of the variables: the codebook command. Either type codebook in the Command window and press *Enter* or navigate the menus to **Data > Describe data > Describe data contents (codebook)**, and click **OK**. We get a large amount of output that is worth investigating. Look it over to see that much can be learned from this simple command. You can scroll back in the Results window to see earlier results, if need be. We'll focus on the output for make, rep78, and foreign.

To start our investigation, we would like to run the codebook command on just one variable, say, make. We can do this via menus or the command line, as usual. To get the codebook output for `make` via the menus, start by navigating as before, to **Data > Describe data > Describe data contents (codebook)**. When the dialog appears, there are multiple ways to tell Stata to consider only the `make` variable:

- We could type `make` into the *Variables* field.
- The *Variables* field is actually a combobox control that accepts variable names. Clicking the button to the right of the *Variables* field displays a list of the variables from the current dataset. Selecting a variable from the list will, in this case, enter the variable name into the edit field.

A much easier solution is to type `codebook make` in the Command window and then press *Enter*. The result is informative:

```
. codebook make

make                                                          Make and Model

                  type:  string (str18), but longest is str17
         unique values:  74                        missing "":  0/74
              examples:  "Cad. Deville"
                         "Dodge Magnum"
                         "Merc. XR-7"
                         "Pont. Catalina"
               warning:  variable has embedded blanks
```

The first line of the output tells us the variable name (`make`) and the variable label (`Make and Model`). The variable is stored as a string (which is another way of saying "text") with a maximum length of 18 characters, though a size of only 17 characters would be enough. All the values are unique, so if need be, `make` could be used as an identifier for the observations—something that is often useful when putting together data from multiple sources or when trying to weed out errors from the dataset. There are no missing values, but there are blanks within the makes. This latter fact could be useful if we were expecting `make` to be a single-word string variable.

Syntax note: telling the `codebook` command to run on the `make` variable is an example of using a *varlist* in Stata's syntax.

Looking at the `foreign` variable can teach us about value labels. We would like to look at the codebook output for this variable, and on the basis of our latest experience, it would be easy to type `codebook foreign` into the Command window to get the following output:

```
. codebook foreign

foreign                                                            Car type

                  type:  numeric (byte)
                 label:  origin

                 range:  [0,1]                           units:  1
         unique values:  2                           missing .:  0/74

            tabulation:  Freq.   Numeric  Label
                            52         0  Domestic
                            22         1  Foreign
```

We can glean that `foreign` is an indicator variable, because its only values are 0 and 1. The variable has a value label that displays "Domestic" instead of 0 and "Foreign" instead of 1. There are two advantages of storing the data in this form:

- Storing the variable as a byte takes less memory, because each observation uses 1 byte instead of the 8 bytes needed to store "Domestic". This is important in large datasets. See [U] **12.2.2 Numeric storage types**.

- As an indicator variable, it is easy to incorporate into statistical models. See [U] **25.2 Using indicator variables in estimation**.

Finally, we can learn a little about a poorly labeled variable with missing values by looking at the rep78 variable. Typing codebook rep78 into the Command window yields

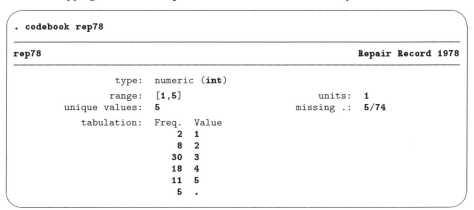

rep78 appears to be a categorical variable, but because of lack of documentation, we do not know what the numbers mean. (To see how we would label the values, see chap. 10.) This variable has five missing values, meaning that there are 5 observations for which the repair record is not recorded. We can use the Data Browser to investigate these 5 observations—and we will do this using the Command window only. (Doing so is much simpler.) If you recall from earlier, the command brought up by clicking the Data Browser button was browse. We would like to browse only those observations for which rep78 is missing, so we could type

```
. browse if missing(rep78)
```

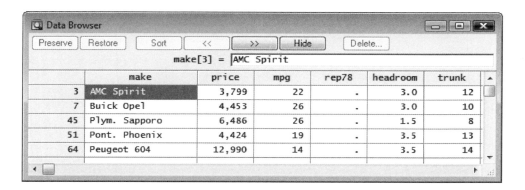

From this, we see that the . entries are indeed missing values—though other missing values are allowable. See [U] **12.2.1 Missing values**. Close the Data Browser after you are satisfied with this statement.

Syntax note: using the *if qualifier* above is what allowed us to look at a subset of the observations.

Looking through the data lends no clues about why these particular data are missing. We decide to check the source of the data to see if the missing values were originally missing or if they were omitted in error. Listing the makes of the cars whose repair records are missing will be all we need, since we saw earlier that the values of make are unique. This can be done via the menus and a dialog.

1. Select **Data > Describe data > List data**.
2. Click the button to the right of the *Variables* field to show the variable names.
3. Click make to enter it into the *Variables* field.
4. Click the **by/if/in** tab in the dialog.
5. Type missing(rep78) into the *If: (expression)* box.
6. Click **Submit**. Stata executes the proper command but the dialog box remains open. **Submit** is useful when experimenting, exploring, or building complex commands. We will use primarily **Submit** in the examples. You may click **OK** in its place if you like.

The same ends could be achieved by typing list make if missing(rep78). The latter is easier, once you know that the command list is used for listing observations. In any case, here is the output:

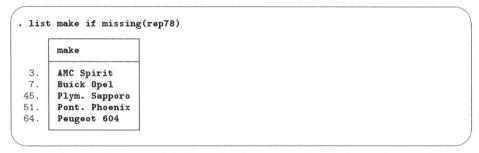

```
. list make if missing(rep78)

         +----------------+
         |  make          |
         |----------------|
   3.    |  AMC Spirit    |
   7.    |  Buick Opel    |
  45.    |  Plym. Sapporo |
  51.    |  Pont. Phoenix |
  64.    |  Peugeot 604   |
         +----------------+
```

We go to the original reference and find that the data were truly missing and cannot be resurrected. See chapter 11 for more information about all that can be done with the list command.

Syntax note: this command uses two new concepts for Stata commands—the if qualifier and the missing() function. The if qualifier restricts the observations on which the command runs to only those observations for which the expression is true. See [U] **11.1.3 if exp**. The missing() function tests each observation to see if it is missing. See [U] **13.3 Functions**.

Now that we have a good idea about the underlying dataset, we can investigate the data themselves.

Descriptive statistics

We saw above that the summarize command gave brief summary statistics about all the variables. Suppose now that we became interested in the prices while summarizing the data, because they seemed fantastically low (it was 1978, after all). To get an in-depth look at the price variable, we can use the menus and a dialog:

1. Select **Statistics > Summaries, tables, and tests > Summary and descriptive statistics > Summary statistics**.
2. Enter or select price in the *Variables* field.

3. Select **Display additional statistics**.

4. Click **Submit**.

Syntax note: as can be seen from the Results window, typing `summarize price, detail` will get the same result. The portion after the comma contains *options* for Stata commands; hence, `detail` is an example of an option.

```
. summarize price, detail
                           Price

            Percentiles      Smallest
  1%            3291            3291
  5%            3748            3299
 10%            3895            3667        Obs               74
 25%            4195            3748        Sum of Wgt.       74

 50%          5006.5                       Mean         6165.257
                               Largest     Std. Dev.    2949.496
 75%            6342           13466
 90%           11385           13594       Variance      8699526
 95%           13466           14500       Skewness     1.653434
 99%           15906           15906       Kurtosis     4.819188
```

From the output, we can see that the median price of the cars in the dataset is only about $5,006! We could also see that the four most expensive cars are all priced between $13,400 and $16,000. If we wished to browse the cars that are the most expensive (and gain some experience with Stata's command syntax), we could use an `if` qualifier,

```
. browse if price > 13000
```

and see from the Data Browser that the expensive cars are two Cadillacs and two Lincolns, which have low gas mileage and are fairly heavy:

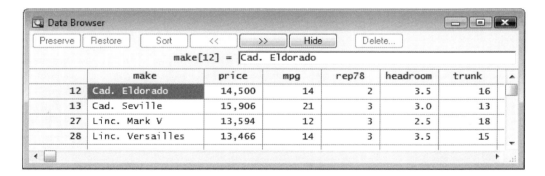

We now decide to turn our attention to foreign cars and repairs, because as we glanced through the data, it appeared that the foreign cars had better repair records. Let's start by looking at the proportion of foreign cars in the dataset along with the proportion of cars with each type of repair record. We can do this with one-way tables. The table for `foreign` cars can be done via menus and a dialog starting with **Statistics > Summaries, tables, and tests > Tables > One-way tables** and then choosing the variable `foreign` in the *Categorical variable* field. Clicking **Submit** yields

```
. tabulate foreign
  Car type |      Freq.      Percent         Cum.
-----------+-----------------------------------
  Domestic |         52        70.27        70.27
   Foreign |         22        29.73       100.00
-----------+-----------------------------------
     Total |         74       100.00
```

We see that roughly 70% of the cars in the dataset are domestic, whereas 30% are foreign made. The value labels are used to make the table, so that the output is nicely readable.

Syntax note: we also see that this one-way table could be made by using the `tabulate` command together with a one variable: `foreign`.

Making a one-way table for the repair records is simple—it will be simpler if done via the Command window. Typing `tabulate rep78` yields

```
. tabulate rep78
     Repair |
Record 1978 |      Freq.      Percent         Cum.
------------+-----------------------------------
          1 |          2         2.90         2.90
          2 |          8        11.59        14.49
          3 |         30        43.48        57.97
          4 |         18        26.09        84.06
          5 |         11        15.94       100.00
------------+-----------------------------------
      Total |         69       100.00
```

We can see that most cars have repair records of 3 and above, though the lack of value labels makes us unsure what a "3" means. The five missing values are indirectly evident, because the total number of observations listed is 69 rather than 74.

These two one-way tables do not help us compare the repair records of foreign and domestic cars. A two-way table would help greatly, which we can get by using the menus and a dialog:

1. Select **Statistics > Summaries, tables, and tests > Tables > Two-way tables with measures of association**.
2. Choose `rep78` as the row variable.
3. Choose `foreign` as the column variable.
4. It would be nice to have the percentages within the `foreign` variable, so check the **Within-row relative frequencies** checkbox.
5. Click **Submit**.

Here is the resulting output.

(Continued on next page)

```
. tabulate rep78 foreign, row
```

```
Key

  frequency
row percentage
```

Repair Record 1978	Car type Domestic	Foreign	Total
1	2	0	2
	100.00	0.00	100.00
2	8	0	8
	100.00	0.00	100.00
3	27	3	30
	90.00	10.00	100.00
4	9	9	18
	50.00	50.00	100.00
5	2	9	11
	18.18	81.82	100.00
Total	48	21	69
	69.57	30.43	100.00

The output indicates that foreign cars are generally much better then domestic cars when it comes to repairs. If you like, you could repeat the previous dialog and try some of the hypothesis tests available from the dialog. We will abstain.

Syntax note: we see that typing the command `tabulate rep78 foreign, row` would have given us the same table. Thus, using `tabulate` with two variables yielded a two-way table. It makes sense that `row` is an option, because we went out of our way to check it in the dialog, so we changed the behavior of the command from its default.

Continuing our exploratory tour of the data, we would like to compare gas mileages between foreign and domestic cars, starting by looking at the summary statistics for each group by itself. We know that a direct way to do this would be to summarize `mpg` for each of the two values of `foreign`, by using `if` qualifiers:

```
. summarize mpg if foreign==0
```

Variable	Obs	Mean	Std. Dev.	Min	Max
mpg	52	19.82692	4.743297	12	34

```
. summarize mpg if foreign==1
```

Variable	Obs	Mean	Std. Dev.	Min	Max
mpg	22	24.77273	6.611187	14	41

It appears that foreign cars get somewhat better gas mileage—we will test this soon.

Syntax note: we needed to use a double equal sign (==) for testing equality. This could be familiar to you if you have programmed before. If it is unfamiliar, it is a common source of errors when initially using Stata. Thinking of equality as "really equal" can cut down on typing errors.

There are two other methods that we could have used to produce these summary statistics. These methods are worth knowing, because they are less error prone. The first method duplicates the concept of what we just did by exploiting Stata's ability run a command on each of a series of nonoverlapping subsets of the dataset. To use the menus and a dialog, do the following:

1. Select **Statistics > Summaries, tables, and tests > Summary and descriptive statistics > Summary statistics** and click the **Reset** button.

2. Select mpg in the *Variables* field.

3. Select *Standard display* option (if it is not already selected).

4. Select the **by/if/in** tab.

5. Check the *Repeat command by groups* checkbox.

6. Select or type foreign for the grouping variable.

7. **Submit** the command.

You can see that the results match those from above. They have a better appearance because the value labels are used rather than the numerical values. The method is more appealing because the results were produced without knowing the possible values of the grouping variable ahead of time.

```
. by foreign, sort: summarize mpg

-> foreign = Domestic
    Variable |        Obs        Mean    Std. Dev.       Min        Max
         mpg |         52    19.82692    4.743297         12         34

-> foreign = Foreign
    Variable |        Obs        Mean    Std. Dev.       Min        Max
         mpg |         22    24.77273    6.611187         14         41
```

There is something different about the equivalent command that appears above: it contains a *prefix command* called a by prefix. The by prefix has its own option, namely, **sort**, to ensure that like members are adjacent to each other before being summarized. The by prefix command is important for understanding data manipulation and working with subpopulations within Stata. Store this example away, and consult [U] **11.1.2 by varlist:** and [U] **27.2 The by construct** for more information. Stata has other prefix commands for specialized treatment of commands, as explained in [U] **11.1.10 Prefix commands**.

The third method for tabulating the differences in gas mileage across the cars' origins involves thinking about the structure of desired output. We need a one-way table of automobile types (foreign versus domestic) within which we see information about gas mileages. Looking through the menus yields the menu item **Statistics > Summaries, tables, and tests > Tables > One/two-way table of summary statistics**. Selecting this, entering foreign for *Variable 1* and mpg for the *Summarize variable*, and submitting the command yields a nice table:

```
. tabulate foreign, summarize(mpg)
                    Summary of Mileage (mpg)
    Car type  |      Mean      Std. Dev.         Freq.

    Domestic  |   19.826923    4.7432972             52
     Foreign  |   24.772727    6.6111869             22

       Total  |   21.297297    5.7855032             74
```

The equivalent command is evidently `tabulate foreign, summarize(mpg)`.

Syntax note: this is a one-way table, so `tabulate` uses one variable. The variable being summarized is passed to the `tabulate` command via an option.

A simple hypothesis test

We would like to run a hypothesis test for the difference in the mean gas mileages. Under the menus, **Statistics > Summaries, tables, and tests > Classical tests of hypotheses > Two-group mean-comparison test** leads to the proper dialog. Enter mpg for the *Variable name* and foreign for the *Group variable name*, and **Submit** the dialog. The results are

```
. ttest mpg, by(foreign)
Two-sample t test with equal variances

    Group |      Obs        Mean     Std. Err.    Std. Dev.    [95% Conf. Interval]

 Domestic |       52    19.82692      .657777     4.743297     18.50638    21.14747
  Foreign |       22    24.77273      1.40951     6.611187     21.84149    27.70396

 combined |       74     21.2973     .6725511     5.785503      19.9569    22.63769

     diff |              -4.945804    1.362162                 -7.661225   -2.230384

     diff = mean(Domestic) - mean(Foreign)                          t =  -3.6308
 Ho: diff = 0                                       degrees of freedom =       72

    Ha: diff < 0                 Ha: diff != 0                  Ha: diff > 0
 Pr(T < t) = 0.0003      Pr(|T| > |t|) = 0.0005        Pr(T > t) = 0.9997
```

From this, we could conclude that the mean gas mileage for foreign cars is different from that of domestic cars (though we really ought to have wanted to test this before snooping through the data). We can also conclude that the command, `ttest mpg, by(foreign)` is easy enough to remember. Feel free to experiment with unequal variances, various approximations to the number of degrees of freedom, and the like.

Syntax note: the `by()` option used here is not the same as the `by` prefix command used earlier. Although it has a similar conceptual meaning, its usage is different because it is a particular option for the `ttest` command.

Descriptive statistics, correlation matrices

We now change our focus from exploring categorical relationships to exploring numerical relationships: we would like to know if there is a correlation between miles per gallon and weight. We select **Statistics > Summaries, tables, and tests > Summary and descriptive statistics > Correlations and covariances** in the menus. Entering `mpg` and `weight`, either by clicking or typing, and then submitting the command yields

```
. correlate mpg weight
(obs=74)

             |      mpg   weight
-------------+------------------
         mpg |   1.0000
      weight |  -0.8072   1.0000
```

The equivalent command for this is natural: `correlate mpg weight`. There is a negative correlation, which is not surprising, because heavier cars should be harder to push about.

We could see how the correlation compares for foreign and domestic cars by using our knowledge of how the by prefix works. We can reuse the *correlate* dialog or use the menus as before if the dialog is closed. Click the **by/if/in** tab, check the *Repeat command by groups* checkbox, and enter the `foreign` variable to define the groups. As done above on page 19, a simple by foreign, sort: prefix in front of our previous command would work, too:

```
. by foreign, sort: correlate mpg weight

-> foreign = Domestic
(obs=52)

             |      mpg   weight
-------------+------------------
         mpg |   1.0000
      weight |  -0.8759   1.0000

-> foreign = Foreign
(obs=22)

             |      mpg   weight
-------------+------------------
         mpg |   1.0000
      weight |  -0.6829   1.0000
```

We see from this that the correlation is not as strong among the foreign cars.

Syntax note: although we used the `correlate` command to look at the correlation of two variables, Stata can make correlation matrices for an arbitrary number of variables:

(Continued on next page)

```
. correlate mpg weight length turn displacement
(obs=74)

                       mpg    weight    length     turn displa~t

         mpg      1.0000
      weight     -0.8072    1.0000
      length     -0.7958    0.9460    1.0000
        turn     -0.7192    0.8574    0.8643    1.0000
displacement     -0.7056    0.8949    0.8351    0.7768    1.0000
```

This can be useful, for example, when investigating collinearity among predictor variables. In fact, simply typing `correlate` will yield the complete correlation matrix.

Graphing data

We have found several things in our investigations so far: We know that the average MPG of domestic and foreign cars differs. We have learned that domestic and foreign cars differ in other ways as well, such as in frequency-of-repair record. We found a negative correlation of MPG and weight—as we would expect—but the correlation appears stronger for domestic cars.

We would now like to examine, with an eye toward modeling, the relationship between MPG and weight, starting with a graph. We can start with a scatterplot of `mpg` against `weight`. The command for this is simple: `scatter mpg weight`. Using the menus requires a few steps because the graphs in Stata can be customized heavily.

1. Select **Graphics > Twoway graph (scatter, line, etc.)**.
2. Click the **Create...** button.
3. Select the *Basic plots* radio button (if it is not already selected).
4. Select *Scatter* as the basic plot type (if it is not already selected).
5. Select `mpg` as the Y *variable* and `weight` as the X *variable*.
6. Click the **Submit** button.

The Results window shows the command that was issued from the menu:

```
. twoway (scatter mpg weight)
```

The command issued when the dialog was submitted is a bit more complex than the command suggested above. There is good reason for this: the more complex structure allows combining and overlaying graphs, as we will soon see. In any case, the graph that appears is

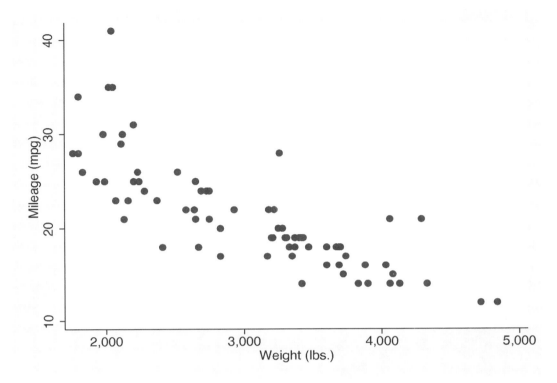

We see the negative correlation in the graph, though the relationship appears to be nonlinear.

Note: when you draw a graph, the Graph window appears, probably covering up your Results window. Click on the main Stata window to get the Results window back on top. Want to see the graph again? Click the **Graph** button. See chapter 15 for more information about the **Graph** button.

We would now like to see how the different correlations for foreign and domestic cars are manifested in scatterplots. It would be nice to see a scatterplot for each type of car, along with a scatterplot for all the data.

Syntax note: since we are looking at subgroups, this looks like it is a job for by. We want one graph—think about whether it should be a by() option or a by prefix.

Start as before:

1. Select **Graphics > Twoway graph (scatter, line, etc.)** from the menus.
2. If *Plot 1* is displayed under the *Plot definitions*, and (scatter mpg weight) appears below the *Plot definitions* box, select it and skip to step 4.
3. Go through the process to create the graph on the previous page.
4. Click the **By** tab.
5. Check the *Draw subgraphs for unique values of variables* checkbox.
6. Enter foreign as the *Variable*.
7. Check the *Add a graph with totals* checkbox.
8. Click the **Submit** button.

The command and the associated graph are

```
. twoway (scatter mpg weight), by(foreign, total)
```

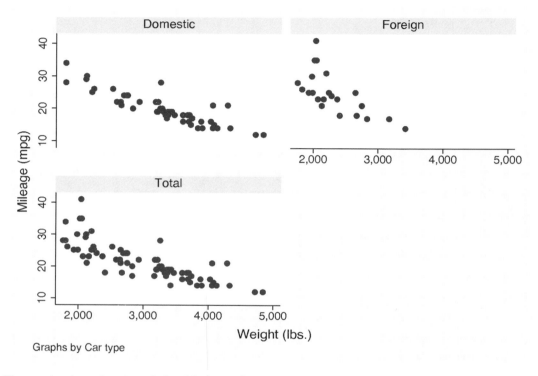

The graphs show that the relationship is nonlinear for both types of cars.

Syntax note: this is another use of a by() option. If you had used a by prefix, two separate graphs would have been generated.

Model fitting: Linear regression

After looking at the graphs, we would like to fit a regression model that predicts MPG from the weight and type of the car. From the graphs, the relationship is nonlinear and so we will try modeling MPG as a quadratic in weight. Also from the graphs, we judge the relationship to be different for domestic and foreign cars. We will include an indicator (dummy) variable for foreign and evaluate afterward whether this adequately describes the difference. Thus, we will fit the model

$$\mathtt{mpg} = \beta_0 + \beta_1 \, \mathtt{weight} + \beta_2 \, \mathtt{weight}^2 + \beta_3 \, \mathtt{foreign} + \epsilon$$

foreign is already an indicator (0/1) variable, but we need to create the weight-squared variable. This can be done via the menus, but here using the command line is simpler:

```
. generate wtsq = weight^2
```

Now that we have all the variables we need, we can run a linear regression. We'll use the menus and see that the command is also simple. To use the menus, select **Statistics > Linear models and related > Linear regression**. In the resulting dialog, choose mpg as the dependent variable and weight, wtsq, and foreign as the independent variables. **Submit** the command. Here is the equivalent simple regress command and the resulting analysis-of-variance table.

```
. regress mpg weight wtsq foreign

      Source |       SS       df       MS              Number of obs =      74
-------------+------------------------------           F(  3,    70) =   52.25
       Model |  1689.15372        3   563.05124         Prob > F      =  0.0000
    Residual |   754.30574       70  10.7757963         R-squared     =  0.6913
-------------+------------------------------           Adj R-squared =  0.6781
       Total |  2443.45946       73  33.4720474         Root MSE      =  3.2827

         mpg |      Coef.   Std. Err.      t    P>|t|     [95% Conf. Interval]
-------------+----------------------------------------------------------------
      weight |  -.0165729   .0039692    -4.18   0.000    -.0244892   -.0086567
        wtsq |   1.59e-06   6.25e-07     2.55   0.013     3.45e-07    2.84e-06
     foreign |    -2.2035   1.059246    -2.08   0.041      -4.3161   -.0909002
       _cons |   56.53884   6.197383     9.12   0.000     44.17855    68.89913
```

The results look encouraging, so we will plot the predicted values on top of the scatterplots for each of the types of cars. To do this, we need the predicted, or fitted, values. This can be done via the menus, but doing it by hand is simple enough. We will create a new variable mpghat:

```
. predict mpghat
(option xb assumed; fitted values)
```

The output from this command is simply a notification. Go over to the Variables window and scroll to the bottom to confirm that there is now an mpghat variable. If you were to try this command when mpghat already exists, Stata will refuse to overwrite your data:

```
. predict mpghat
mpghat already defined
r(110);
```

The predict command, when used after a regression, is called a *postestimation command*. As specified, it creates a new variable called mpghat equal to

$$-.0165729 \, \mathtt{weight} + 1.59 \times 10^{-6} \, \mathtt{wtsq} - 2.2035 \, \mathtt{foreign} + 56.53884$$

For careful model fitting, there are several features available to you after estimation—one is calculating predicted values. Be sure to read [U] **20 Estimation and postestimation commands**.

We can now graph the data and the predicted curve to evaluate the fit on the foreign and domestic data separately to determine if our shift parameter is adequate. We can draw both graphs together. Using the menus and a dialog, do the following:

1. Select **Graphics > Twoway graph (scatter, line, etc.)** from the menus.
2. If there are any plots listed, click the **Reset** button, 🅡.
3. Create the graph for `mpg` versus `weight`.
    a. Click the **Create...** button.
    b. Be sure that *Basic plots* and *Scatter* are selected.
    c. Select `mpg` as the *Y variable* and `weight` as the *X variable*.
    d. Click **Accept**.
4. Create the graph showing `mpghat` versus `weight`.
    a. Click the **Create...** button.
    b. Be sure that *Basic plots* and *Line* are selected.
    c. Select `mpghat` as the *Y variable* and `weight` as the *X variable*.
    d. Check the *Sort on x variable* box. Doing so ensures that the lines connect from smallest to largest `weight` values, instead of the order in which the data happen to be.
    e. Click **Accept**.
5. Show two plots, one each for domestic and foreign cars, on the same graph.
    a. Click the **By** tab.
    b. Check the *Draw subgraphs for unique values of variables*.
    c. Enter `foreign` in the *Variables* field.
6. Click the **Submit** button.

Here are the resulting command and graph.

```
. twoway (scatter mpg weight) (line mpghat weight, sort), by(foreign)
```

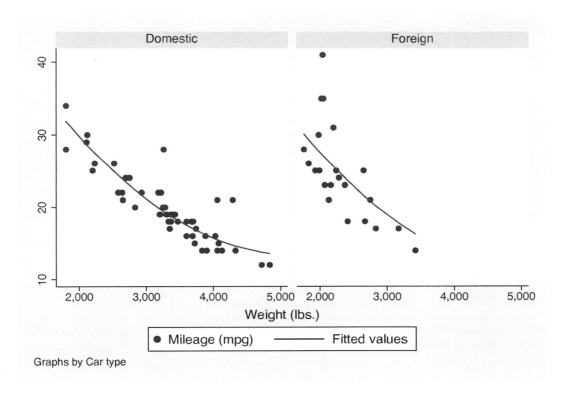

Graphs by Car type

Here we can see the reason for enclosing the separate `scatter` and `line` commands in parentheses: they can then be overlaid by submitting them together. The fit of the plots looks good and is cause for initial excitement. So much excitement, in fact, that we decide to print the graph and show it to an engineering friend. We print the graph, being careful to print the graph and not all our results: **File > Print** from the Graph window menu bar.

When we show our graph to our engineering friend, she is concerned. "No," she says. "It should take twice as much energy to move 2,000 pounds 1 mile compared with moving 1,000 pounds the same distance, and therefore it should consume twice as much gasoline. Miles per gallon is not a quadratic in weight; gallons per mile is a linear function of weight. Don't you remember any physics?"

We try out what she says. We need to generate a gallons-per-mile variable and make a scatterplot. Here are the commands that we would need—note their similarity to commands issued earlier in the session. There is one new command: the `label variable` command, which allows us to give the gpm variable a *variable label* so that the graph is labeled nicely.

```
. generate gpm = 1/mpg
. label variable gpm "Gallons per Mile"
. twoway (scatter gpm weight), by(foreign, total)
```

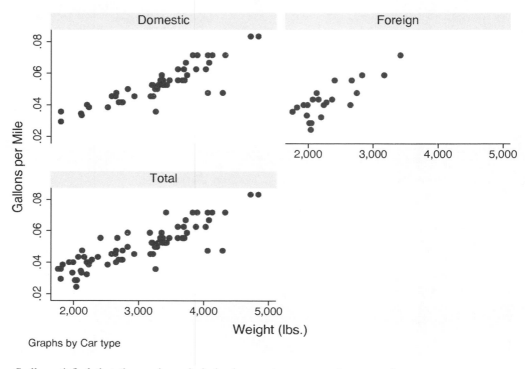

Graphs by Car type

Sadly satisfied that the engineer is indeed correct, we rerun the regression:

```
. regress gpm weight foreign
```

Source	SS	df	MS		Number of obs =	74
					F(2, 71) =	113.97
Model	.009117618	2	.004558809		Prob > F =	0.0000
Residual	.00284001	71	.00004		R-squared =	0.7625
					Adj R-squared =	0.7558
Total	.011957628	73	.000163803		Root MSE =	.00632

gpm	Coef.	Std. Err.	t	P>|t|	[95% Conf. Interval]	
weight	.0000163	1.18e-06	13.74	0.000	.0000139	.0000186
foreign	.0062205	.0019974	3.11	0.003	.0022379	.0102032
_cons	-.0007348	.0040199	-0.18	0.855	-.0087504	.0072807

We find that although foreign cars had better gas mileage than domestic cars in 1978, it was because they were so light. In fact, according to our model a foreign car with the same weight as a domestic car would use an additional 1/160 gallon per mile driven. With this, we are satisfied with our analysis.

Commands versus menus

In this chapter you have seen that Stata can operate either via menu choices and dialog boxes or via the Command window. As you become more familiar with Stata, you will find that the Command window is typically much faster for often-used commands, whereas the menus and dialogs are faster when building up complex commands such as graphs.

One of Stata's great strengths is the consistency of its *command syntax*. Most of Stata's commands share the following syntax, where square brackets mean that something is optional and a *varlist* is a list of variables.

$$\left[\textit{prefix:} \right] \textit{ command } \left[\textit{varlist} \right] \left[\textit{if} \right] \left[\textit{in} \right] \left[\textit{weight} \right] \left[\textit{, options} \right]$$

Some general rules:

- Most commands accept prefix commands that modify their behavior; see [U] **11.1.10 Prefix commands** for details. One of the more common prefix commands is by.
- If an optional *varlist* is not specified, all the variables are used.
- *if* and *in* restrict the observations on which the command is run.
- *options* modify what the command does.
- Each command's syntax is found in the online help and the reference manuals.
- Stata's command syntax includes more than we have shown you here, but this should get you started. For more information, see [U] **11 Language syntax** and help language.

We saw examples using all the pieces of this except for the in qualifier and the weight clause. The syntax for all commands can be found in the online help along with examples—see chapter 6 for more information. The explanation of the syntax can be found online using help language. The consistent syntax makes it straightforward to learn new commands and to read others' commands when examining an analysis.

Here is an example of reading the syntax diagram by using the summarize command from earlier in this chapter. The syntax diagram for summarize is typical:

$$\texttt{summarize} \left[\textit{varlist} \right] \left[\textit{if} \right] \left[\textit{in} \right] \left[\textit{weight} \right] \left[\textit{, options} \right]$$

This means that

command by itself is valid:	summarize
command followed by a *varlist* (variable list) is valid:	summarize mpg
	summarize mpg weight
command with *if* (with or without a *varlist*) is valid:	summarize if mpg>20
	summarize mpg weight if mpg>20

and so on.

You can learn about summarize in [R] **summarize**, or select **Help > Stata Command...** and enter summarize, or type help summarize in the Command window.

Keeping track of your work

It would have been useful if we had made a log of what we did so that we could conveniently look back at interesting results or track any changes that were made. You will learn to do this in chapter 17. Your logs will contain commands and their output—another reason to learn command syntax, so that you can remember what you've done.

To make a log file that keeps track of everything appearing in the Results window, click the button that looks like a lab notebook, . Choose a place to store your log file, and give it a name, just as you would any other document. The log file will save everything that appears in the Results window from the time you start a log file to the time that you close it.

Conclusion

This chapter introduced you to Stata's capabilities. You should now read and work through the rest of this manual. Once you are done here, you can read the *User's Guide*.

4 The Stata user interface

The windows

This chapter introduces the core of Stata's interface: its main windows, its toolbar, its menus, and its dialog boxes.

Past commands appear here   Variable list appears here   Results are displayed here

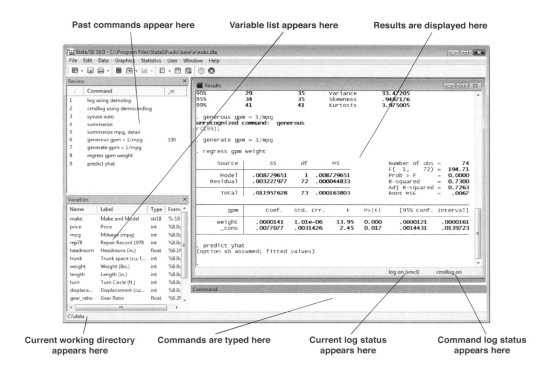

Current working directory   Commands are typed here   Current log status   Command log status
appears here   appears here   appears here

The four main windows are the Review, Variables, Results, and Command windows. These are initially contained in the main Stata window. Each has its name in its title bar. These four windows are typically in use the whole time Stata is open. There are other, more specialized windows, such as the Viewer, Data Editor, Data Browser, Do-file Editor, Graph, and Graph Editor windows—these are discussed later in this manual.

To open any window, or to reveal a window hidden by other windows, select the window from the **Window** menu or select the proper item from the toolbar. Many of Stata's windows have functionality that can be accessed by clicking the right mouse button (right-clicking) on the window. Right-clicking displays a contextual menu that, depending on the window, allows you to copy text, set the preferences for the window, or print the contents of the window. When copying text or printing, we recommend that you always right-click on the window rather than use the menu bar or toolbar so that you can be sure of where and what you're copying or printing.

The toolbar

This is the toolbar:

The toolbar contains buttons that provide quick access to Stata's more commonly used features. If you forget what a button does, hold the mouse pointer over a button for a moment and a tooltip will appear with a description of that button. Buttons that contain both an icon and an arrow display a menu if you click the arrow.

Open: Open a Stata dataset. Click the button to open a dataset with the **Open** dialog. Hold down the button to select a dataset from a menu of recently opened datasets.

Save: Save the Stata dataset currently in memory to disk.

Print: Print Results window. Click on the arrow to select a window to print.

Log: Begin a new log or close, suspend, or resume the current log. See chapter 17 for an explanation of log files.

Viewer: Open the Viewer, or bring a Viewer to the front of all other windows. Click the button to open a new Viewer. Click on the arrow to select a Viewer to bring to the front. See chapter 5 for more information.

Graph: Bring a Graph window to the front of all other windows. Click the button to bring the topmost Graph window to the front. Click on the arrow to select a Graph window to bring to the front. See chapter 15 for more information.

Do-file Editor: Open the Do-file Editor, or bring a Do-file Editor to the front of all other windows. Click the button to open a new Do-file Editor. Click on the arrow to select a Do-file Editor to bring to the front. See chapter 14 for more information.

Data Editor: Open the Data Editor, or bring the Data Editor to the front of the other Stata windows. See chapter 8 for more information.

Data Browser: Open the Data Browser, or bring the Data Browser to the front of the other Stata windows. See chapter 8 for more information.

Clear —more— Condition: Tell Stata to continue when it has paused in the middle of long output. See chapter 11 for more information.

Break: Stop the current task in Stata. See chapter 11 for more information.

The Command window

Commands are submitted to Stata from the Command window. The Command window supports basic text editing, copying and pasting, a command history, function-key mapping, and variable-name completion. From the Command window, pressing

Page Up	Steps backward through the command history.
Page Down	Steps forward through the command history.
Tab	Auto-completes a partially typed variable name, if possible.

See [U] **10 Keyboard use** for more information about keyboard shortcuts for the Command window.

The command history allows you to recall a previously submitted command, edit it if you wish, and then resubmit it. Commands submitted by Stata's dialogs are also included in the command history, so you can recall and submit a command without having to open the dialog again.

The Review window

The Review window shows the history of commands that have been entered, displaying successful commands in black, and unsuccessful commands, along with their error codes, in red. To enter a command from the Review window,

- Click once on a past command to copy it to the Command window, replacing the contents of the Command window.
- Double-click on a past command to execute it. (Executing the command adds the command to the bottom of the Review window, also.)

Right-clicking on the Review window displays a menu from which you can select

- **Cut** to remove the selected command(s) from the Review window and place them on the Clipboard.
- **Copy** to copy the selected command(s) to the Clipboard.
- **Delete** to remove the selected command(s) from the Review window.
- **Select All** to select all the commands in the Review window, including those before and after the commands currently displayed.
- **Clear All** to clear out all the commands from the Review window, including those before and after the commands currently displayed.
- **Do Selected** to submit all the selected commands and add them to the bottom of the command history. Stata will attempt to run all the selected commands, even those containing errors, and will not stop even if a command causes an error.
- **Send to Do-file Editor** to place all the selected commands into a new Do-file Editor window.
- **Save All...** to bring up a *Save Review Contents* dialog, allowing you to save all the commands in the Review window, including those before and after the commands currently displayed, in a *do-file*. (See chap. 14 for more information on do-files.)
- **Save Selected...** to bring up a *Save Review Contents* dialog, allowing you to save the selected commands in the Review window in a do-file.
- **Font...** to bring up a *Font* dialog, allowing you to change the font used to display the Review window contents.

The Variables window

The Variables window shows the list of variables, along with their labels, storage types, and display formats, for the dataset in memory. Click once on a variable in the Variable window to paste it onto the end of whatever is in the Command window. Double-clicking a variable is like clicking on the variable once. Right-clicking on a variable in the Variables window displays a menu from which you can select

- **Rename '***varname***'...** to rename the variable *varname*.
- **Edit Variable Label for '***varname***'...** to change the variable label for *varname*.
- **Format '***varname***'...** to change the display format for *varname*.
- **Drop '***varname***'...** to drop, or eliminate, *varname* from the dataset in memory. You will be asked for confirmation. This affects only the dataset in memory, not the dataset as saved on your disk. See chapter 7 for more information.

- **Define Notes for** *'varname'*... add, edit, or drop notes attached to variable *varname*. See chapter 10 for more information.

- **Define Notes for Data...** add, edit, or drop notes attached to the dataset. See chapter 10 for more information.

- **Font...** brings up a *Font* dialog, allowing you to change the font used to display the Variables window contents.

Menus and dialogs

There are two ways by which you can tell Stata what you would like it to do: you can use menus and dialogs, or you can use the Command window. When you worked through the sample session in chapter 3, you saw that both have strengths. We will discuss the menus and dialogs here.

Stata's **Data**, **Graphics**, and **Statistics** menus provide point-and-click access to almost every command in Stata. As you will learn, Stata is fully programmable, and Stata programmers can even create their own dialogs and menus. The **User** menu provides a place for programmers to add their own menu items. Initially, it contains only some empty submenus.

If you wish to perform a Poisson regression, for example, you could type Stata's poisson command, or you could select **Statistics > Count outcomes > Poisson regression**, which would display this dialog:

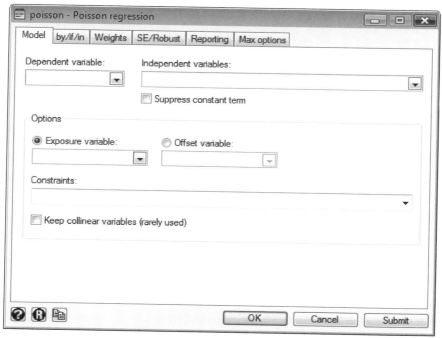

This dialog provides access to all of the functionality of Stata's poisson command. The poisson command has many options that can be accessed by clicking the multiple tabs across the top of the dialog. The first time you use the dialog for a command, it is a good idea to look at the contents of each tab so that you will know all the dialog's capabilities.

The dialogs for many commands have the **by/if/in** and **Weights** tabs. These provide access to Stata's commands and qualifiers for controlling the estimation sample and dealing with weighted data. See [U] **11 Language syntax** for more information on these features of Stata's language.

The dialogs for most estimation commands have the **Max options** tab for setting the maximization options (see [R] **maximize**). For example, you can specify the maximum number of iterations for the optimizer.

Most dialogs in Stata provide the same six buttons you see at the bottom of the Poisson dialog on the previous page.

 OK issues a Stata command based on how you have filled out the fields in the dialog and then closes the dialog.

 Cancel closes the dialog without doing anything—just as clicking on the red close button does.

 Submit issues a command just like **OK** but leaves the dialog on the screen so that you can make changes and issue another command. This feature is handy when, for example, learning a new command or putting together a complicated graph.

 Help provides access to Stata's help system. Clicking this button will typically take you to the help file for the Stata command associated with the dialog. Here, clicking on it would take you to the `poisson` help file. The help file will have tabs above groups of options to show which dialog tab contains those options.

 Reset resets the dialog to its default state. Each time you open a dialog, it will remember how you last filled it out. If you wish to reset its fields to their default values at any time, simply click this button.

Copy Command to Clipboard behaves much like the **Submit** button, but rather than issuing a command, it copies the command to the Clipboard. The command can then be pasted elsewhere (such as the Do-file Editor).

The command issued by a dialog is submitted just as if you typed it by hand. You can see it in the Results window and the Review window after it executes. Looking carefully at the full command will help you learn Stata's command syntax.

In addition to being able to access the dialogs for Stata commands through Stata's menus, you can also invoke them by using two other methods. You may know the name of a Stata command for which you want to see a dialog, but you may not remember how to navigate to that command in the menu system. Simply type `db` *commandname* to launch the dialog for *commandname*:

```
. db poisson
```

You will also find access to the dialog for a command in that command's help file; see chapter 6 for more details.

As you read this manual, we will present examples of Stata commands. You may type those examples as presented, but you should also experiment with submitting those commands by using their dialogs. Use the `db` command described above to quickly launch the dialog for any command that you see in this manual.

The working directory

If you look at the screen shot on page 31 you will notice the status bar at the base of the main Stata window that contains C:\data. This indicates that C:\data is the current working directory. The working directory is the folder where graphs and datasets will be saved when typing commands such as save *filename*. It does not affect the behavior of menu-driven file actions such as **File > Save** or **File > Open...**. Once you have started Stata, you can change the current working directory with the cd command. See [D] **cd** for full details. Stata always displays the name of the working directory so that it is easy to tell where your graphs and datasets will be saved.

Fine control of Stata's windows

When Stata is first launched, the Review, Variables, Results, and Command windows appear within the main Stata window. The Review, Variables, and Command windows are initially linked to each other and attached to the edges of the Stata window. They cannot be moved, though they can be resized by dragging on their edges.

The Results window initially can be moved around inside the main Stata window but cannot be moved outside the main Stata window. It can be maximized, which then locks it into whatever space remains inside the main Stata window.

All other windows that open up can be moved freely and independently of the Stata window.

This default behavior is simple and familiar, but it can be modified to suit your work style and the size of your screen. The following sections explain how you can adjust window behavior to fit your fancy.

Window types

The Stata for Windows interface has two types of windows: *docking* and *nondocking*. The Review, Variables, Command, and Viewer windows are docking windows. All other windows are nondocking.

Docking windows are those that have special characteristics that allow them to be used with other docking windows: they can be *linked* (sharing a window with a splitter dividing the windows) or *tabbed* (sharing a window with a tab for each). They can be set to auto-hide when not needed. Nondocking windows have none of these special characteristics.

Here is brief comparison between the two types of windows:

Docking windows
- can be dragged in and out of the main Stata window or other docking windows,
- can be linked to other docking windows so that they share a window with a splitter dividing them,
- can share one window with other docking windows with a tab for each, and
- can be made to hide automatically when not in use.

Nondocking windows
- are separate from the main Stata window (except for the Results window),
- cannot be linked to other windows,
- cannot be tabbed to another window,
- cannot be made to hide automatically when not in use, and
- can be made to always appear in front of the main Stata window.

When Stata is first launched, the Review, Variables, Results, and Command windows appear within the main Stata window. The Review, Variables, and Command windows are docking windows and are initially both docked and linked to each other. The Results window is a nondocking window.

Docking windows

To allow docking windows to have extra features, you first need to select **Edit > Preferences > General Preferences...**, click on the **Windowing** tab, and check **Enable ability to dock, undock, or tab windows**. Until this is selected, the Review, Variables, and Command windows will remain docked on the edges of the main Stata window.

Docking windows can be dragged into (docked) and out of (undocked) the main Stata window or other docking windows. When you drag a docking window, a transparent blue box appears in place of the window. If you drag the box over the main Stata window or another docking window, docking guides appear (see figure 1). Dragging the box over a guide previews how and where the window will appear if you release the mouse button. Releasing the mouse button when the box is over a guide (dropping) links or tabs the window as previewed, whereas releasing it when the box is not on a docking guide simply moves the window to that position.

Dropping the box on one of the outer docking guides links the docking window to the window under the guide. For example, figure 1 shows how you would drag the Review window over the Variables window, at which point the docking guides appear. When you move the mouse over the bottom docking guide and release the mouse button, the Review window becomes linked to the Variables window and is displayed in the lower part of the window.

When docking windows are linked, a divider (splitter) is inserted between them. The splitter allows the linked windows to be resized: increasing the size of one window decreases the size of the other. To prevent splitters from resizing linked windows, select **Edit > Preferences > General Preferences...**, click on the **Windowing** tab, and check **Lock splitters**. To undock a linked window, double-click on the window's title bar, or drag the window away from the other window.

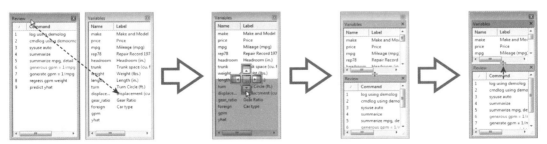

Figure 1: Linking two windows

Dropping the box on the center docking guide (see figure 2) combines the docking window with the window under the guide. One docking window is created with a tab for each docked window at the bottom of the new window. Clicking on a tab displays its window. To undock a tabbed window, double-click on the window's tab, or click and drag the window's tab out of the docking window. The docking window containing the tabbed windows can also be linked to other docking windows.

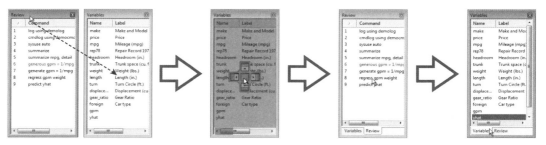

Figure 2: Docking two windows

You can temporarily disable the docking guides by holding down the *Ctrl* key while dragging a docking window. To permanently disable the docking guides, select **Edit > Preferences > General Preferences...**, click on the **Windowing** tab, and uncheck *Use docking guides*.

Auto Hide and pinning

Docking windows support the Auto Hide feature, which automatically hides the docking windows when they are not in use. This feature is turned off by default. To enable this feature, select **Edit > Preferences > General Preferences...**, click on the **Windowing** tab, and check *Enable ability to pin or unpin windows*. When this feature is enabled, a pushpin will appear in the title bar of all docked Docking windows.

To enable Auto Hide for a docking window, click the pushpin in the title bar (see figure 3) so that the pushpin points left. Click the pushpin again to disable Auto Hide; the pushpin will point down.

After you enable Auto Hide for a docking window, the window is hidden when not in use and is displayed as a tab along an edge of the main Stata window. To show the window, move the mouse pointer over the tab. To hide the window again, move the mouse pointer outside the window. If the window has the focus (its title bar is highlighted), it will be hidden if you click on another window. Auto Hide works only on docking windows that are docked within the main Stata window.

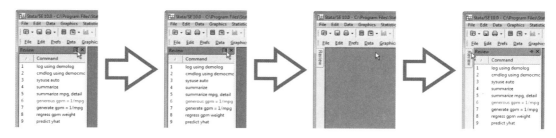

Figure 3: Enabling Auto Hide

When you undock a docking window, you can move it in front of or behind other Stata windows, including the main Stata window. If you click on any part of the main Stata window, the main window will move to the front and obscure the undocked window. You can bring the undocked window back to the front by selecting it from the **Window** menu. Or you can just make undocked windows always appear in front of the main Stata window (float) by selecting **Edit > Preferences > General Preferences...**, clicking on the **Windowing** tab, and checking *Make windows float*.

In addition to dragging the window, you can dock or undock a docking window by double-clicking its title bar. You can double-click the title bar again to move the window back to its previously docked or undocked location.

If you enable Auto Hide for the Review and Variables windows in their default docked location, they will obscure any text at the beginning of the Command window when they are displayed. To avoid this problem, dock these two windows at the right edge of the main Stata window before enabling Auto Hide.

Nondocking windows

Nondocking windows can be dragged in and out of the main Stata window but do not share any of the other special features of docking windows, such as linking and tabbing. Nondocking windows include the Graph window, Data Editor, Do-file Editor, and dialogs. The Results window, because of its status as the primary window of Stata, must remain inside the main Stata window.

As with docking windows, the main Stata window, when brought to the front, can obscure nondocking windows that exist outside the main window. To prevent this, select **Edit > Preferences > General Preferences...**, click on the **Windowing** tab, and check *Make windows float*. This setting makes both undocked docking windows and nondocking windows always appear in front.

Making windows float is useful if you like to run the main Stata window maximized. However, windows that are floating can obscure part of the Results window because they always appear on top of the main Stata window. You can avoid this problem by selecting **Edit > Preferences > General Preferences...**, clicking on the **Windowing** tab, and checking *Make Results window float*. This setting allows the Results window to exist outside the main Stata window and float in front of it. You can move other windows in front of or behind the Results window when it is floating.

Notes

5 Using the Viewer

The Viewer's purpose

The Viewer is a versatile tool in Stata. It will be the first place you can turn for help within Stata, but it is far more than just a help system. You can also use the Viewer to keep your copy of Stata current; add, delete, and manage third-party extensions to Stata known as *user-written programs*; view and print Stata logs both from your current and previous Stata sessions; view and print any other Stata-formatted (SMCL) or plain-text (ASCII) file; and even launch your browser to follow hyperlinks.

This chapter focuses on the general use of the Viewer, its buttons, and a brief summary of the commands that the Viewer understands. There is more information about using the Viewer to find help in chapter 6 and for maintaining your copy of Stata in chapter 20.

To open a new Viewer window, you can either click the **Viewer** button, , or select **Window > Viewer > New Viewer**. A Viewer opened in this fashion provides links that allow you to perform several tasks.

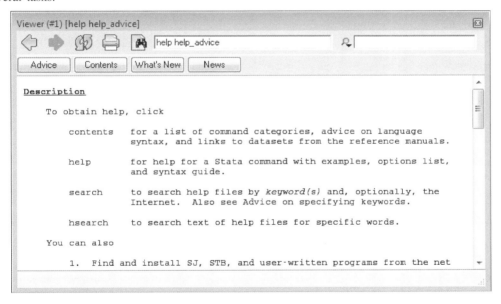

Viewer buttons

The toolbar of the Viewer has multiple buttons, a command box, and a search box.

 Back: Goes back one step in your viewing trail.

Forward: Goes forward one step in your viewing trail, assuming you had backtracked.

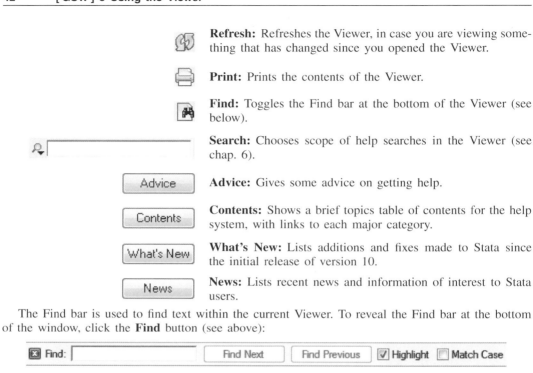

Refresh: Refreshes the Viewer, in case you are viewing something that has changed since you opened the Viewer.

Print: Prints the contents of the Viewer.

Find: Toggles the Find bar at the bottom of the Viewer (see below).

Search: Chooses scope of help searches in the Viewer (see chap. 6).

Advice: Gives some advice on getting help.

Contents: Shows a brief topics table of contents for the help system, with links to each major category.

What's New: Lists additions and fixes made to Stata since the initial release of version 10.

News: Lists recent news and information of interest to Stata users.

The Find bar is used to find text within the current Viewer. To reveal the Find bar at the bottom of the window, click the **Find** button (see above):

The Find bar has its own buttons, fields, and checkboxes.

Close: Closes the Find bar.

Find: The field for entering the *search text* you would like to find.

Find Next: Jumps to the next instance of the search text. Automatically wraps past the end of the Viewer document if there are no further instances of the search text.

Find Previous: Jumps to the previous instance of the search text. Automatically wraps past the start of the Viewer document if there are no previous instances of the search text.

Highlight: If checked, highlights other instances of the search text (in yellow, by default). If unchecked, only the current instance of the search text is highlighted (in black, by default). By default, this is checked.

Match Case: If checked, uppercase and lowercase letters are considered different, so searching for This would not find this. If unchecked, uppercase and lowercase letters are considered the same, so searching for This would find this. By default, this box is unchecked.

Viewer's function

The Viewer is similar to a web browser. It has clickable links (shown in blue text) that you can follow to related help topics, install and manage third-party software, and even keep your copy of Stata up to date. When you move the mouse pointer over a link, the status bar at the bottom of the Viewer shows the action associated with that link. If the action of a link is `help logistic`, clicking that link will show the help file for the `logistic` command in the Viewer. Middle-clicking on a link in a Viewer window (if you do not have a three-button mouse, shift-clicking) will open the link in a new Viewer window.

You can open a new Viewer by selecting **Window > Viewer > New Viewer** or clicking the Viewer button on the toolbar, or by right-clicking on an existing Viewer window and selecting **Open New Viewer**. Entering a `help` command from the Command window will also open a new Viewer.

To bring a Viewer to the front of all other Viewers, select **Window > Viewer** and choose a Viewer from the list there. Selecting **Close All Viewers** closes all open Viewer windows.

The Viewer is a docking window (see chap. 4). A Viewer that is docked with the main Stata window when Stata is exited will be opened automatically the next time Stata is started. To also make undocked Viewers automatically open, select **Edit > Preferences > General Preferences...**, click the **Windowing** tab, and check *Make undocked Viewers persistent*.

A Viewer that is automatically opened initially displays the Viewer help file. You can make it display the same contents it displayed when it was last open by selecting **Edit > Preferences > General Preferences...**, clicking on the **Windowing** tab, and checking *Persistent Viewers retain topics*.

Viewing local text files, including SMCL files

In addition to viewing built-in Stata help files, you can use the Viewer to view Stata Markup and Control Language (SMCL) files, such as those typically produced when logging your work (see chap. 17) and plain-text files. To open a file and view its contents, simply select **File > View...**, and you will be presented with with a dialog:

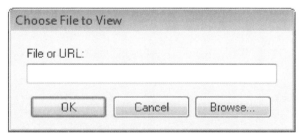

You may either type in the name of the file that you wish to view and click **OK**, or you may click the **Browse...** button to open a standard file dialog that allows you to navigate to the file.

If you currently have a log file open, you may view the log file in the Viewer. This method has one advantage over scrolling back in the Results window: what you view stays fixed even as output is added to the Results window. If you wish to view a current log file, select **File > Log > View...**, and the usual dialog will appear, but with the path and filename of the current log already in the field. Simply click **OK**, and the log will appear in the Viewer. See chapter 17 for more details.

Viewing remote files over the Internet

If you want to look at a remote file over the Internet, the process is similar to viewing a local file, only instead of using the **Browse...** button, you type in the URL of the file that you want to see, such as http://www.stata.com/man/readme.smcl. You should use the Viewer only to view text or SMCL files. If you enter the URL of, say, an arbitrary web page, you will see the HTML source of the page instead of the usual browser rendering.

Navigating within the Viewer

In addition to using the window scrollbar to navigate the Viewer window, you also can use the up/down cursor keys and *Page Up/Page Down* keys to do the same. Pressing the up/down cursor keys scrolls the window a line at a time. Pressing the *Page Up/Page Down* keys scrolls the window a screenful at a time.

Printing

To print the contents of the Viewer, right-click on the window, and select **Print...**. You may also select **File > Print >** *"viewer name"* or click on the arrow the **Print** toolbar button to select from a menu of open windows to print.

Right-clicking on the Viewer window

Right-clicking on the Viewer window displays a contextual menu from which you can do many of the tasks listed above.

- **Back** to go back one step in your viewing trail.
- **Forward** to go forward one step in your viewing trail, assuming you had backtracked.
- **Open New Viewer** to open a new Viewer.
- **Open Link in New Viewer** to open a link in a new Viewer (as long as you right-clicked on the link itself).
- **Close All Viewers But This One** to close all other Viewer windows.
- **Copy** to copy highlighted text to the Clipboard.
- **Copy Table** to copy a table to the Clipboard in tab-delimited format.
- **Copy Table as HTML** to copy a table to the Clipboard in HTML format, i.e., a format usable on a web page. Copying a nontable as HTML is the same as using the **Copy** menu item.
- **Copy as Picture** to copy selected text in the Viewer window as a picture in WMF format.
- **Select All** to select all text in the Viewer.
- **Find in Viewer...** to find text in the Viewer by using the Find bar.
- **Preferences...** to edit the preferences for the Viewer window.
- **Font...** to change the font for the window.
- **Print...** to print the contents of the Viewer window.

Searching for help in the Viewer

The *search box* in the Viewer can be used to search documentation. Click the magnifying glass; choose **Search documentation and FAQs**, **Search net resources**, or **Search all**; and then type a word or phrase in the search window and press *Enter*. For more extensive information about using the Viewer for help, see chapter 6.

Commands in the Viewer

Everything you can do in the Viewer by clicking on links and buttons can also be done by typing commands in the *command box* at the top of the window or on the Stata command line. Some of the commands that can be issued in the Viewer are

1. *Obtaining help* (see chap. 6)

    Type `contents` to view the contents of Stata's help system.

    Type *commandname* to view the help file for a Stata command.

2. *Searching* (see chap. 6)

    Type `search` *keyword* to search documentation and FAQs on a topic.

    Type `search` *keyword*`, net` to search net resources on a topic.

    Type `search` *keyword*`, all` to search both of the above.

3. *User-written programs* (see chap. 6 and chap. 20)

    Type `net from http://www.stata.com/` to find and install *Stata Journal*, STB, and user-written programs from the net.

    Type `ado` to review user-written programs you have installed.

    Type `ado uninstall` to uninstall user-written programs you have installed on your computer.

4. *Updating* (see chap. 20)

    Type `update` to check your current Stata version.

    Type `update query` to check for new official Stata update releases.

    Type `update all` to update your Stata.

5. *Viewing files in the Viewer*

    Type `view` *filename*`.smcl` to view SMCL files.

    Type `view` *filename*`.txt` to view ASCII files.

    Type `view` *filename*`.log` to view ASCII log files.

6. *Viewing files in the Results window*

    Type `type` *filename*`.smcl` to view SMCL files in the Command window.

    Type `type` *filename*`.txt` to view ASCII files in the Command window.

    Type `type` *filename*`.log` to view ASCII log files in the Command window.

7. *Launching your browser to view an HTML file*

    Type `browse` *URL* to launch your browser.

8. *Keeping informed*

    Type `news` to see the latest news from http://www.stata.com.

(Continued on next page)

Using the Viewer from the Command window

Stata allows you to print Viewer windows from the Command window. Each Viewer has a unique name (which is displayed in parentheses in the window's title bar) to help identify it from the command line. For example, the name of a Viewer with the window title *Viewer (#1) [help regress]* is **#1**. Its contents can be printed by typing the command `print @Viewer, name(#1)`. Printing in this way does not bring up a print dialog.

Typing `help` *commandname* in the Command window will bring up a new Viewer showing the requested help.

6 Getting help

Online help

Stata's help system provides a wealth of information to help you learn and use Stata. To find out which Stata command will perform the statistical or data-management task you would like to do, you should generally follow these steps:

1. Select **Help > Search...**, choose **Search documentation and FAQs**, and enter the topic or keyword(s). This search will find information about Stata commands, references to articles in the *Stata Journal* or the *Stata Technical Bulletin* (STB), links to FAQs on Stata's web site, and links to selected external web sites.

2. Read through the **Search** results, and click on the appropriate command name link to open its help file.

3. Read the help file for the command you chose.

4. If the first help file you went to is not what you wanted, either look at the end of the help file for links to take you to related help files or click the **Back** button to go back to the previous document and go from there to other help files.

5. With the help file open, click on the Command window and enter the command, or click on one of the dialog links at the top right of the help file to open a dialog for the command.

6. If, at any time, you want to begin again with a new **Search**, enter the new search terms in the search box of the Viewer window.

7. If your **Search documentation and FAQs** search returned no results, you can look for *Stata Journal*, STB, and user-written programs on the Internet (materials available via Stata's `net` command) by entering your search term(s) in the search box, clicking the magnifying glass, and selecting **Search net resources**.

8. If you select **Search documentation and FAQs**, Stata searches Stata's keyword database. If you select **Search net resources**, Stata searches for *Stata Journal*, STB, and user-written programs available for free download on the Internet; see chapter 20 for more information.

9. You can also select **Search all**, which is equivalent to choosing both **Search documentation and FAQs** and **Search net resources**. This is the equivalent of Stata's `findit` command.

Let's illustrate the **Help** system with an example. You'll get the most from the example if you work along at your computer.

Suppose that we have been given a dataset about antique cars, and we need to know what it contains. Though we still have a vague notion of having seen something like this while working through the example session in chapter 3, we do not remember the proper command.

Start by typing `sysuse auto` in the Command window to bring the dataset into memory.

Following the above approach, we

1. Select **Help > Search...**.

2. Check that the **Search documentation and FAQs** radio button is selected.

3. Enter `dataset contents` into the search box, and click **OK** or press *Enter*. Before pressing *Enter*, the window should look like

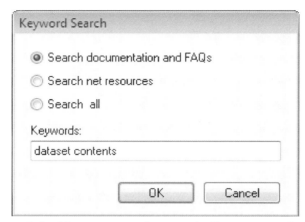

4. Stata will now search for "dataset contents" among the Stata commands, the reference manuals, the *User's Guide*, the *Stata Journal*, the *Stata Technical Bulletin*, and the FAQs on Stata's web site. Here is the result:

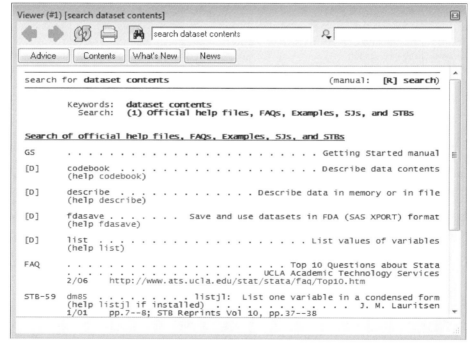

5. Upon seeing the results of the search, we see two commands that look promising: codebook and describe. Since we are interested in the contents of the dataset, we decide to check out the codebook command. The [D] means that we could look up the codebook command in the *Stata Data Management Reference Manual*. The blue codebook link in (help codebook) means that there is an online help file for the codebook command. This is what we are interested in right now.

6. Click the blue codebook link. Links can go to a variety of resources, such as help for Stata commands, dialog boxes, and even web pages. Here the link goes to the help file for the codebook command.

7. What is displayed is typical for help for a Stata command. Help files for Stata commands contain, from top to bottom,

     a. Links to dialog(s) that can be useful in issuing the command. In this example, there is one link to the `codebook` dialog.

     b. The command's syntax, i.e., rules for constructing a command that Stata will correctly interpret. The square brackets here indicate that all the arguments to `codebook` are optional but that if we wanted to specify them, we could use a *varlist*, an `if` qualifier, or an `in` qualifier, along with some options. (Options vary greatly from command to command.) The options are listed directly under the command and are explained in some detail later in the help file. You will learn more about command syntax in chapter 11.

     c. A description of the command. Since "codebook" is the name for big binders containing hardcopy describing each of the elements of a dataset, the description for the `codebook` command is justifiably terse.

     d. The options that can be used with this command. These are explained in much greater detail than in the listing of the possible options after the syntax. Here, for example, we can see that the `mv` option can look to see if there is a pattern in the missing values—something important for data cleaning and imputation.

     e. Examples of command usage. The `codebook` examples are real examples that step through using the command on a dataset either shipped with Stata or loadable within Stata from the Internet.

     f. References to related commands. A useful list of cross-references.

For now, scroll down to the examples. It is worth going through the examples as given in the help file. Here is a screen shot of the top of the examples:

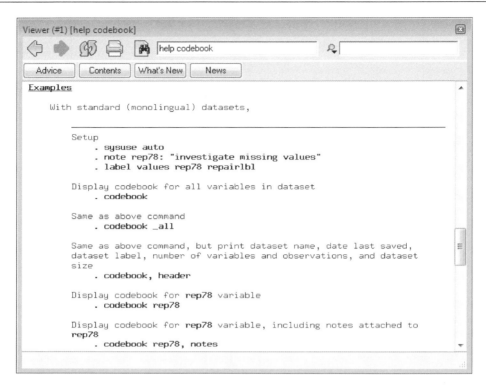

Searching help

Search is designed to help you find information about statistics, graphics, data management, and programming features in Stata. When entering topics for the search, use appropriate terms from statistics, etc. For example, you could enter Mann-Whitney. Multiple topic words are allowed, e.g., regression residuals. For advice on using **Search**, click the **Advice** button, and follow your choice of advice links.

When you are using **Search**, use proper English and proper statistical terminology. If you already know the name of the Stata command and want to go directly to its help file, select **Help > Stata Command...**, and type the command name. You can also type the command name in the search field at the top of the Viewer and press *Enter*.

Help distinguishes between topics and Stata commands, because some names of Stata commands are also general topic names. For example, logistic is a Stata command. If you choose **Stata Command...** and enter logistic, you will go right to the help file for the command. But, if you choose **Search...** and enter logistic, you will get search results listing the many Stata commands that relate to logistic regression.

Remember that you can search for help from within a Viewer window by typing a command in the Command box of the Viewer or by selecting the scope of the search by using the button to the left of the Search box, typing the search criteria in the Search box, and pressing *Enter*.

Contents

If you click the **Contents** button, you will see several help categories.

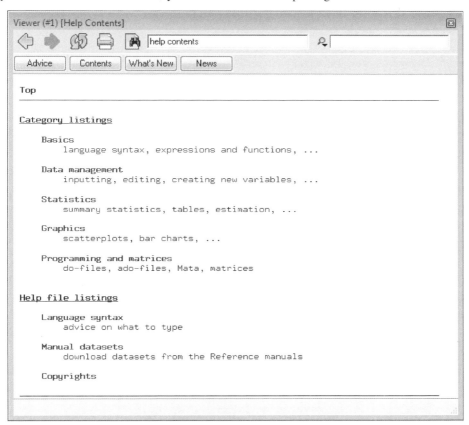

Click on a link to display more information about that category. For example, if you click on **Statistics**, more information about statistics will be displayed.

Help and search commands

As you may expect, the help system is accessible from the Command window. This feature is especially convenient when you need help on a particular Stata command. Here is a short listing of the various commands you can use:

1. Typing `search` *topic* in the Command window produces the same output as selecting **Help > Search...**, choosing **Search documentation and FAQs**, and entering *topic*. The output appears in the Results window, not in a Viewer.

2. Typing `search` *topic*, `net` in the Command window produces the same output as selecting **Help > Search...**, choosing **Search net resources**, and entering *topic*. The output appears in the Results window, not in a Viewer.

3. Typing `search` *topic*, `all` in the Command window produces the same output as selecting **Help > Search...**, choosing **Search all**, and entering *topic*. The output appears in the Results window, not in a Viewer.

4. Typing `help` *commandname* is equivalent to selecting **Help > Stata Command...** and entering *commandname*. The help file for the command appears in a new Viewer.

5. Typing `chelp` *commandname* is similar to selecting **Help > Stata Command...** and entering *commandname*, except the help file for the command appears in the Results window instead of a Viewer.

See [U] **4 Stata's online help and search facilities** and [U] **4.8 search: All the details** in the *User's Guide* for more information about these command language versions of the **Help** system. The `search` command, in particular, has a few capabilities (such as author searches) that we have not demonstrated here.

The Stata reference manuals and User's Guide

Notations such as [R] **ci**, [R] **regress**, and [R] **ttest** in the **Search** results and help files are references to the three-volume *Stata Base Reference Manual*. You may also see things like [P] **#delimit**, which is a reference to the *Stata Programming Reference Manual*, and [U] **9 The Break key**, which is a reference to the *Stata User's Guide*. For a complete list of manuals and their shorthand notations, see *Cross-Referencing the Documentation*, which immediately follows the table of contents in this manual. Why bother to look in the manual when help is available in Stata? The manual has more information. Although Stata's interactive help is extensive, it contains only about 1/10 of the information contained in the reference manuals.

The Stata reference manuals are each arranged like an encyclopedia, alphabetically, and each has its own index. The *User's Guide* also has an index. The *Quick Reference and Index* contains a combined index for the *User's Guide* and all the reference manuals. This combined index is a good place to start when you are looking for information about a command.

Entries have names like **collapse**, **egen**, ..., **summarize**, which are generally themselves Stata commands.

For advice on how to use the reference manuals, see chapter 19, or see [U] **1.1 Getting Started with Stata**.

The Stata Journal and the Stata Technical Bulletin

The **Search documentation and FAQs** facility searches all the Stata source materials, including the online help, the *User's Guide*, the reference manuals, this manual, the *Stata Journal*, the *Stata Technical Bulletin* (STB), and the FAQs on Stata's web site.

The **Search net resources** facility searches all materials available via Stata's `net` command; see [R] **net**. These materials include the *Stata Journal*, STB, and user-written additions to Stata available on the Internet.

The *Stata Journal* is a printed and electronic journal, published quarterly, containing articles about statistics, data analysis, teaching methods, and effective use of Stata's language. The *Journal* publishes reviewed papers together with shorter notes and comments, regular columns, tips, book reviews, and other material of interest to researchers applying statistics in a variety of disciplines. The *Journal* is a publication for all Stata users, both novice and experienced, with different levels of expertise in statistics, research design, data management, graphics, reporting of results, and of Stata, in particular. See http://www.stata-journal.com/ for more information.

The predecessor to the *Stata Journal* was the *Stata Technical Bulletin* (STB). Even though the STB is no longer published, past issues contain articles and programs that may interest you. See http://www.stata.com/bookstore/stbj.html for the table of contents of past issues, and see the STB FAQ at http://www.stata.com/support/faqs/res/stb.html for detailed information on the STB.

Associated with each issue of both the *Stata Journal* and the STB are the programs and datasets described therein. These programs and datasets are made available for download and installation over the Internet, not only to subscribers, but to all Stata users. See [R] **net** and [R] **sj** for more information.

Since the *Stata Journal* and the STB have had several articles dealing with regression models, if you select **Help > Search...** and choose **Search documentation and FAQs**, and enter ordered regression, you will see some of these references:

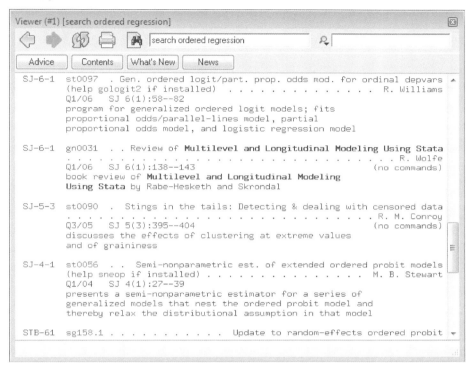

SJ-6-1 refers to volume 6, number 1 of the *Stata Journal*. Abbreviations like st0097 and gn0031 are the way the *Stata Journal* refers to the articles that have appeared in it.

STB-61 refers to the 61st issue of the STB. Every six issues make a year's worth of STBs, and they are bound and republished in a volume called *Stata Technical Bulletin Reprints*. STB-59 could thus be found in *STB Reprints*, volume 11. Abbreviations like sg158.1 and sg163 are the way the STB refers to the articles that have appeared in it.

In the screen shot above, you will notice that article st0097 in SJ-6-1 has a command, gologit2, associated with it. You can install this command by using Stata's Viewer. You may also install commands from STB articles via the Internet. See chapter 20 for more information.

Links to other sites where you can freely download programs and datasets for Stata can be found on the Stata web site; see http://www.stata.com/links/. See chapter 20 for more details on how to install this software. Also see [R] **ssc** for information on a convenient interface to resources available from the Statistical Software Components (SSC) archive.

We recommend that all users subscribe to the *Stata Journal*. See [U] **3.5 The Stata Journal and the Stata Technical Bulletin** for more information.

Notes

7   Opening and saving Stata datasets

How to load your dataset from disk and save it to disk

Opening and saving datasets in Stata works in fashion similar to that in other computer applications. There are a few differences, however. First off, it is possible to save and open files from within Stata's Command window. Second, Stata allows just one dataset to be open and in use at any one time. It is possible to have many Viewers viewing many files, but only one dataset may be in use at any time. Keeping this in mind will make Stata's care in opening new datasets clear. This chapter outlines all the possible ways to open and save datasets.

A Stata dataset can be opened in a variety of ways, most of which are probably familiar to you from other applications:

- Double-click a Stata data file, which is a file whose extension is dta. **Note:** the file extension may not be visible, depending on what options you have set in your operating system.
- Select **File > Open...** or click the **Open** button and navigate to the file.
- Select **File > Open Recent >** *filename*.
- Type use *filename* in the Command window. Stata will look for *filename* in the working directory. If the file is located elsewhere, you will need to give its path.

Because Stata has at most one dataset open at a time, opening a dataset will cause Stata to discard the dataset that is currently in memory. If there have been changes to the data in the currently open dataset, Stata will refuse to discard the dataset unless you force it to do so. If you open the file with any method other than the Command window, you will be prompted. If you use the Command window, and the current data have changed, you will get the following error:

```
. use something_new
no; data in memory would be lost
r(4);
```

These behaviors protect you from mistakenly losing data.

To save an unnamed dataset (or an old dataset under a new name):

- Select **File > Save As...**.
- Type save *filename* in the Command window.

To save a dataset for use with Stata 8 or Stata 9:

- Select **File > Save As...**, and select **Stata 9 Data (*.dta)** from the **Save as type** list.
- Type saveold *filename* in the Command window.

To save a dataset that has been changed (overwriting the original data file):

- Select **File > Save**.
- Click the **Save** button.
- Simply type save, replace in the Command window.

Once you overwrite a dataset, there is **no** way to recover your original dataset. With important datasets, you may want to either keep a backup copy of your original *filename*.dta or save your changes to a dataset under a new name. This is no different from working with a word-processing document, except that recovering from an inadvertent save with a dataset is nearly impossible.

Important note: changes you have made to a dataset are not permanent until you save them. You work with a copy of the dataset in memory, not with the data file itself. This should not be surprising, because it is the way that you work with most all applications on your computer.

Troubleshooting

If you try to open a dataset and Stata complains that there is not enough memory, your dataset is larger than the amount of memory that Stata is using. This can happen because Stata must load the entire dataset into memory before it can be used. (By default, Stata/MP and Stata/SE use 10 MB of memory for data, and Stata/IC uses 1 MB of memory.) Changing the amount of memory Stata uses can be done easily from within Stata. The memory for Small Stata cannot be changed. For full details, see appendix B of this manual.

8 Using the Data Editor

The Data Editor

The Data Editor gives a spreadsheet-like view on data, if any, that are currently in memory. You can use it to enter new data, edit existing data, and edit attributes of the data in the dataset, such as variable names, labels, and display formats, as well as value labels.

If you are interested only in viewing, but not changing, the dataset, you can open the Data Browser. The Data Browser functions identically to the Data Editor except that it will not allow you to change any data in your dataset.

Buttons on the Data Editor

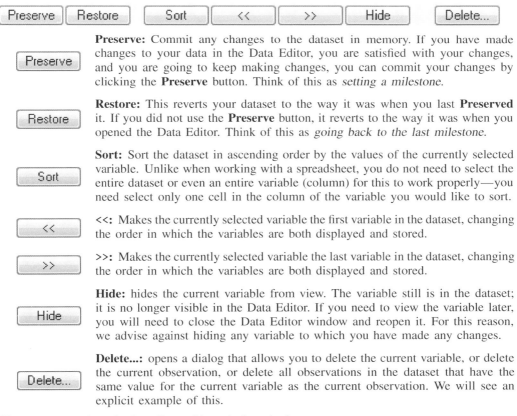

Preserve: Commit any changes to the dataset in memory. If you have made changes to your data in the Data Editor, you are satisfied with your changes, and you are going to keep making changes, you can commit your changes by clicking the **Preserve** button. Think of this as *setting a milestone*.

Restore: This reverts your dataset to the way it was when you last **Preserved** it. If you did not use the **Preserve** button, it reverts to the way it was when you opened the Data Editor. Think of this as *going back to the last milestone*.

Sort: Sort the dataset in ascending order by the values of the currently selected variable. Unlike when working with a spreadsheet, you do not need to select the entire dataset or even an entire variable (column) for this to work properly—you need select only one cell in the column of the variable you would like to sort.

<<: Makes the currently selected variable the first variable in the dataset, changing the order in which the variables are both displayed and stored.

>>: Makes the currently selected variable the last variable in the dataset, changing the order in which the variables are both displayed and stored.

Hide: hides the current variable from view. The variable still is in the dataset; it is no longer visible in the Data Editor. If you need to view the variable later, you will need to close the Data Editor window and reopen it. For this reason, we advise against hiding any variable to which you have made any changes.

Delete...: opens a dialog that allows you to delete the current variable, or delete the current observation, or delete all observations in the dataset that have the same value for the current variable as the current observation. We will see an explicit example of this.

You can move about in the editor with typical methods:
- To move to the right, use the *Tab* key or the right-arrow key.
- To move to the left, use *Shift-Tab* or the left-arrow key.
- To move down, use *Enter* or the down-arrow key.

- To move up, use *Shift-Enter* or the up-arrow key.

Clicking in a cell works also.

Data entry

Entering data into the Data Editor is similar to entering data into a spreadsheet. We'll illustrate this with an example. It will be useful for you to follow the example at your computer. You will need to start with an empty dataset to work along, so save your dataset if necessary, and then type `clear` in the Command window.

Note: as a check to see if your data have changed, type `describe, short` (or `d,s` for short). Stata will tell you if your data have changed.

Suppose that we have the following dataset:

Make	Price	MPG	Weight	Gear Ratio
VW Rabbit	4697	25	1930	3.78
Olds 98	8814	21	4060	2.41
Chev. Monza	3667		2750	2.73
AMC Concord	4099	22	2930	3.58
Datsun 510	5079	24	2280	3.54
	5189	20	3280	2.93
Datsun 810	8129	21	2750	3.55

We do not know MPG for the third car or the make of the sixth.

Start by opening the Data Editor. You can do this by either clicking the **Data Editor** button, 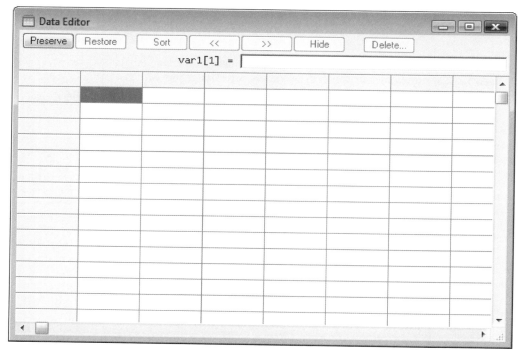, or typing `edit` in the Command window. You should be greeted by a Data Editor with no data displayed.

Stata shows which cell it is working in by highlighting the cell and displaying *varname*[*obsnum*] next to the input box. (The latter can be handy if you scroll the active cell off the screen.) The Data Editor starts, by default, in the first row of the first column. Since there are no data, there are no variable names, and so Stata shows `var1[1]` as the active cell.

We can enter these data either by working across the rows (observation by observation) or by working down the columns (variable by variable). To enter the data observation by observation, press *Tab* after entering each value until you have reached the end of the first row. In our case, we would type `VW Rabbit`, press *Tab*, type `4697`, press *Tab*, and continue entering data to complete the first observation.

After you are finished with the first observation, select the second cell in the first column either by clicking in it or by navigating to it. At this point, your screen should look like this:

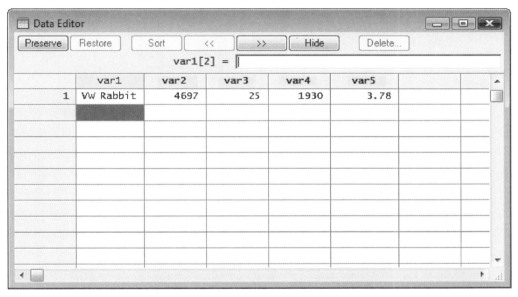

We can now enter the data for the second observation in the same fashion as the first—with one nice difference: after entering the last value in the row, pressing the *Tab* key will bring us to the first cell in the third row. This is possible because the number of variables is known after the first observation has been entered, so Stata knows when it has all the data for an observation.

We can enter the rest of the data by entering data and pressing the *Tab* key between entries, simply skipping over missing values by tabbing through them.

If we had wanted to enter the data variable by variable, we could have done that by pressing *Enter* between each make of car until all 7 observations were entered, skipping past the missing entry by using *Enter* twice in succession. Once the first variable was entered we would select the first cell in the second column and enter the price data. We would continue this until we were done.

(Continued on next page)

Notes on data entry

There are several things to note about data entry and the feedback you get from the Data Editor as you enter data:

1. *Stata does not allow empty columns or rows in the middle of your dataset.*
   Whenever you enter new variables or observations, always begin in the first empty column or row. If you skip over some columns or rows, Stata will fill in the intervening columns or rows with missing values.

2. *Strings and value labels are color coded.*
   To help distinguish between the different types of variables in the editor, string values are displayed in red, value labels (see chap. 10) are displayed in blue, and all other values are displayed in black. You can change the colors for strings and value labels by right-clicking on the editor window and selecting **Preferences...**.

3. *A period ('.') represents Stata's system missing numeric value.*

4. *The Tab key is smart.*
   As we saw above, after the first observation has been entered, Stata knows how many variables you have. So, at the end of the second observation (and all subsequent observations), *Tab* will automatically take you back to the first column.

5. *When you see, for example,* var3[4]=
   This corresponds to the current cell that is highlighted. var3 is the default name of the third variable, and [4] indicates the fourth observation.

6. *Quotes around text are unnecessary in string variables.*
   Once Stata knows that a variable is a string variable (it holds text) there is no need to put quotes around the values, even if the values look like a number. Thus, if you wanted to enter ZIP codes as text, you would enter the first ZIP code with quotes ("02173"), but the rest would not need any quotes.

7. You must use *Tab* or *Enter* when entering or editing data. If you change a data value and use an arrow key to leave the cell, the data will not be changed.

Renaming variables

The data have now been entered into Stata, but the variable names leave something to be desired: they have the default names var1, var2, ..., var5. We would like to rename the variables so that they match the column titles from our dataset.

To rename a variable, start by double-clicking anywhere in the variable's column. This opens the *Variable Properties* dialog where we can change the name, variable label, display format, and value label. (For variable and value labels, see chap. 10; for display formats, see [U] **12.5 Formats: controlling how data are displayed**.) Once we have made the changes we would like, clicking the **OK** button will save them and close the *Variable Properties* dialog. We can now name the first variable make, the second price, the third mpg, the fourth weight, and the fifth gear_ratio. Just before renaming var5 to gear_ratio, your screen should look like this:

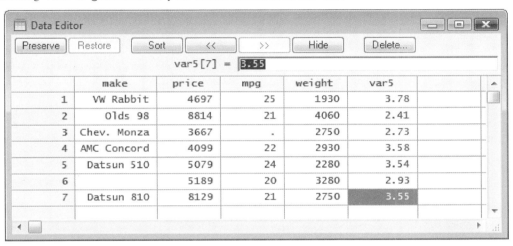

There are some rules for variable names that you need to know:

1. *Stata is case sensitive.*
   Make, make, and MAKE are all different names to Stata. If you had named your variables Make, Price, MPG, etc., you would have to type them correctly capitalized in the future. Using all lowercase letters is easier.
2. *A variable name must be 1–32 characters long.*
3. *The characters can be letters (A–Z, a–z), digits (0–9), or underscores (_).*
4. *Spaces or other characters are not allowed.*
5. *The first character of a variable name must be a letter or an underscore.*
   Although you can use an underscore to begin a variable name, it is highly discouraged. Such names are used for temporary variable names in Stata, but you would like your data to be permanent, so using a temporary name could lead to great frustration.

(Continued on next page)

Copying and pasting data

You can copy and paste data by using the Data Editor. This is often a simple way to bring data into Stata from any other applications such as spreadsheets or databases.

1. *Select the data that you wish to copy.*

    a. Click once on a variable name or column heading to select an entire column.

    b. Click once on an observation number or row heading to select the entire row.

    c. Drag the mouse to select a range of cells.

2. *Copy the data to the Clipboard.*

    Right-click in the selected range, and select **Copy**.

3. *Paste the data from the Clipboard.*

    a. Click on the top left cell of the area to which you wish to paste.

    b. Right-click on the same cell, and select **Paste**.

We will illustrate copying and pasting an observation by making a copy of the first observation and pasting it at the end of the dataset.

Start by clicking on the observation number of the first observation. Doing so highlights all the data in the row. Right-click in the same location (there is no need to move the mouse), and select **Copy**:

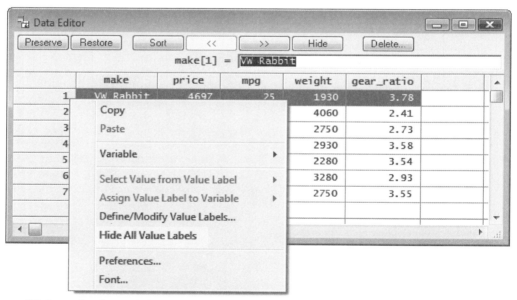

Click on the first cell in the eighth row, right-click right away, and choose **Paste** from the resulting menu. You can see that the observation was successfully duplicated.

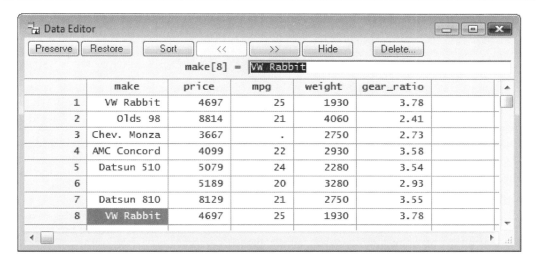

Notes on copying and pasting

1. The above example illustrated copying and pasting within the Data Editor. You can use roughly the same technique to copy and paste between other applications and Stata, and between Stata and other applications. The one requirement for things to work well is that the external application must copy tables either in tab-delimited or comma-delimited form, as do spreadsheet applications, many database applications, and some word processors. The easiest way to see if copying and pasting works properly is to try it. For more information on file-based methods for importing data into Stata, see chapter 9.

2. If you are copying and pasting data with value labels, you have a choice. You can copy variables with value labels as text, using the value labels as the actual values, or you can copy said variables as their underlying encoded numbers. Copying with the value labels is the default. If you would like the other choice, start the Data Editor by entering `edit, nolabel` in the Command window.

(Continued on next page)

Changing data

As its name suggests, the Data Editor can be used to edit your dataset. As we've seen already, it can be used to edit the data themselves as well as the description and display options for the variables. Here is an example for making some changes to the `auto` dataset, which illustrates both methods for using the Data Editor and its documentation trail. We would like to investigate the dataset, pare it down to only foreign cars, and delete the `trunk` variable.

Start by typing `sysuse auto` into the Command window. You may be told that your data have changed—if so, save or clear the dataset in memory. Once the dataset is loaded, start the Data Editor.

1. We would like to see which cars have the lowest and highest gas mileages. To do this, click anywhere in the `mpg` column, and then click the **Sort**. You will see that the data have now been sorted by `mpg` in ascending order. The lowest-mileage cars are at the top of the screen; by scrolling to the bottom of the dataset, you can find the highest-mileage cars.

2. We would like to investigate repair records, and hence sort by the `rep78` variable. (Do this now.) When we scroll to the bottom to see which cars had good repair records, we notice all five of the missing values for `rep78`. A few items of note:

   a. As we can see from the result of the **Sort**, Stata views missing values as being larger than all numeric nonmissing values: `rep78 >= .` is equivalent to `missing(rep78)`.

   b. What we do not see here is that Stata has multiple missing-value indicators: '.' is Stata's default or system missing-value indicator, and `.a`, `.b`, ..., `.z` are Stata's extended missing values. Some people use extended missing values to indicate why a certain value is unknown. Other people have no use for extended missing values and just use '.'.

   c. The different missing values sort within themselves: . < .a < .b < ··· < .z. See [U] **12.2.1 Missing values** for full details.

3. We would like to work with just the foreign cars. To do this, we need to delete all the nonforeign cars from the dataset. Scroll to the right until you find the `foreign` variable. Click on a cell that displays `Domestic`, and then click the **Delete...** button. You will see three choices:

(You may see a different observation number.) By default, Stata offers to delete all observations for which `foreign` matches our selection. Since this is what we would like, click the **OK** button. All the domestic cars have now vanished from the dataset.

4. We are happy with our editing so far, so we set a milestone by clicking the **Preserve** button. By doing this, we are committing ourselves to the changes that we've made, at least for the dataset in memory. Remember that the dataset on the disk is unaffected unless we explicitly save the dataset from Stata.

5. We would now like to delete the `trunk` variable. We will do this by first making a mistake to illustrate what the **Restore** button does. Click anywhere in the `headroom` variable "by mistake", and then click the **Delete** button. Choose to delete the `headroom` variable, and click **OK.** Now you realize that you didn't mean to delete the `headroom` variable. Click the **Restore** button and click the **Yes** button on the **Restore** dialog. Relief: you see that the `headroom` variable has been returned. The **Restore** button will put the dataset back into the state it was in when

it was last **Preserve**d, or back to its state when entering the Data Editor if the **Preserve** button had not been used. This means that it is a good idea to **Preserve** your data from time to time when editing so that a future mistake does not mean a lot of backtracking.

6. We can now drop the **trunk** variable properly. You can do so without further instruction.

7. Since we are done, we can exit the Data Editor. This is done by clicking on the Data Editor window's close box. We made changes since the last **Restore**, so we get a dialog box asking us to confirm the changes. Click **OK**, and the dataset is now ready to go.

8. Your editing commands are listed in the **Results** window so that you can see what you have done. The dash in front of the command indicates that the change was done in the editor.

```
. edit
- preserve
- sort mpg
- sort rep78
- drop if foreign == 0
- preserve
- drop headroom
- restore
- drop trunk
```

9. If we were going to save this dataset, we would have saved it under a new name by using **File > Save As...** so as not to overwrite the original dataset. Since we do not really want these data anymore, we will clear them out of memory by typing `clear` in the Command window.

Data Editor advice

As you could see in the last section, a small mistake in the Data Editor could have large negative ramifications for your dataset. You really must take care in how you edit your data.

1. People who care about data integrity know that editors are dangerous—it is easy to accidentally make changes. **Never use the Data Editor (or the `edit` command) when you just want to look at your data.** Use the Data Browser (or the `browse` command).

2. If you must edit your data, protect yourself by limiting the dataset's exposure. For example, if you need to change `rep78` only if it is missing, find a way to look at just the missing values for `rep78` and any other variables needed to make the change. This will make it impossible for you to change (damage) variables or observations other than those you view. We will explore this aspect shortly.

3. Even with these caveats, Stata's editor is safer than most because it records changes in the Results window. Use this feature to log your output and make a permanent record of the changes. Then, you can verify that the changes you made are the changes you wanted to make. See chapter 17 for information on creating log files.

edit with in and if

We would now like to investigate restricting our view of the data we see in the editor. This feature is useful for the reasons mentioned above, and as we will see, it helps if we would like to browse through the data of a large dataset. In any case, we would like to focus on *some*, not *all* of the data, be it some of the variables, some of the observations, or even just some observations within some variables. This is how it is done.

To use the Data Editor with selected variables and restricted observations, we must use the `edit` command together with a *varlist* (variable list) along with `if` and `in` qualifiers in the Command window. By using a *varlist*, we restrict the variables we look at, whereas the `if` and `in` qualifiers restrict the observations we see. (Chap. 11 contains many examples of using a command with a variable list and `if` and `in`.) As we saw in chapter 3, `make` can be used as a unique identifier of the automobiles. Thus, if we wanted to change the missing values for `rep78`, all that we would need to identify the proper observations would be `make`. To minimize our exposure to making mistakes, we enter the command

```
. sysuse auto
(1978 Automobile Data)
. edit make rep78 if missing(rep78)
```

and we would see following screen:

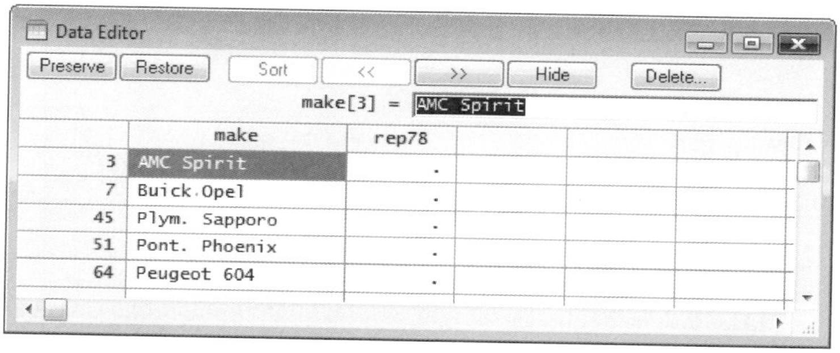

You can see from this that we have done a good job at limiting the damage we can do: at worst we could damage the `make` of one of the cars. We cannot even use the **Sort** button, because we are looking at a subset of the observations. We are safe and sound.

Keep this lesson in mind if you edit your data. It is a lesson well learned.

browse

The purpose of the Data Browser is to look at data without altering it. You can start the Data Browser by clicking the Data Browser button (as was done in chap. 3) or by typing `browse` into the Command window. The Data Browser functions exactly like the Data Editor, except that you may not alter the data, the labels, or any of the display formats for the variables. Just as in the Data Editor, you can restrict your view to a subset of the variables and/or observations, and just as in the Data Editor, if you use an `if` or `in` qualifier, you may not sort the data.

Here is an example of restricting the view. We must use the Command window to browse in such a fashion:

```
. browse make mpg price if foreign==1
```

Using this command results in this window:

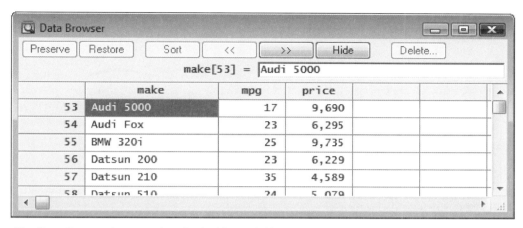

The Data Browser is convenient for looking quickly at large dataset. If you need to have results that you can print nicely, you will need the `list` command, which is covered extensively in chapter 11.

Right-clicking on the Data Editor window

Right-clicking on the Data Editor window displays a menu from which you can do many common tasks:

- **Copy** to copy data to the Clipboard.
- **Paste** to paste data from the Clipboard.
- **Variable > Properties...** to display the **Variable Properties** dialog for the current variable.
- **Variable > Sort** to sort the current variable.
- **Variable > Hide** to hide the current variable.
- **Variable > Delete...** to open the **Data Editor Delete Options** dialog.
- **Select Value from Value Label** '*name*' to display a menu of values for the value label *name*. Select an item from the menu to change the value for the current cell.
- **Assign Value Label to Variable** '*varname*' to display a menu of value labels. Select a value label from the menu to assign it to the current variable.
- **Define/Modify Value Labels...** to display the *Define value labels* dialog.
- **Hide All Value Labels** to hide all value labels and display their numeric value instead.
- **Preferences...** to set the preferences for the editor.
- **Font** to change the font of the editor.

Closing the Data Editor

The following notes also apply to the Data Browser. The Data Editor is *modal*, meaning that while you are editing data, you may not do anything else in Stata. For this, the Command window disappears as soon as you enter the Data Editor. If you find yourself in a situation where the Command window has disappeared and nothing seems to function anymore, your Data Editor has probably moved behind your other windows. You can get it back by selecting **Window > Data Editor** (even if you are using the Data Browser). You can exit the Data Editor by clicking on its close box.

Remember also that any changes made in the Data Editor are made only to the dataset in memory. If you want to save the changes permanently, you must save the dataset after you leave the editor.

Notes

9 Importing data

Copying and pasting

One of the easiest ways to get data into Stata is often overlooked: you can copy data from most applications that understand the concept of a *table* and then paste the data into the Data Editor. This approach works for all spreadsheet applications, many database applications, some word-processing applications, but few web browsers. Just copy the full range of data, paste it into the Data Editor, and everything will probably work well. You can even copy a text file that has the pieces of data separated by commas and then paste it into the Data Editor. There is one warning, however: be sure that your spreadsheet does not contain blank rows, blank columns, repeated headers, or merged cells: these can cause trouble. As long as your spreadsheet looks like a table, you will be fine.

If you cannot simply copy and paste a dataset you would like to use, this chapter is for you. If you would like to learn methods that lend themselves better to repetitive tasks, this chapter is also for you. If your friend has interesting data in a spreadsheet application that you do not own, this chapter is for you.

Commands for importing data

This chapter explains the various commands that Stata has for importing data. The three main commands for reading plain text (ASCII) are

- `insheet`, which is made for reading text files created by spreadsheet or database programs;
- `infile`, which is made for reading simple data that are separated by spaces, or rigidly formatted data aligned in columns; and
- `infix`, which is made for data aligned in columns but possibly split across rows.

Stata has other commands that can read other types of files and can even get data from external databases without the need for an interim file:

- The `fdause` command can read any SAS XPORT file, so data can be transferred from SAS to Stata in this fashion.
- The `odbc` command can be used to pull data directly from any data sources for which you have ODBC drivers.
- The `xmluse` command can read some XML files (most notably Microsoft Excel's SpreadsheetML).

Each command expects the file that they are reading to be in a specific format. This chapter will explain some of those formats and give some examples. For the full story, consult the *Stata Data Management Reference Manual*.

(Continued on next page)

The insheet command

The `insheet` command was specially developed to read in text (ASCII) files that were created by spreadsheet programs. Many people have data that they enter into their spreadsheet programs. All spreadsheet programs have an option to save the dataset as a text (ASCII) file with the columns delimited with either tab characters or commas. Some spreadsheet programs also save the column titles (variable names in Stata) in the text file.

To read in this file, you have only to type `insheet using` *filename*, where *filename* is the name of the text file. The `insheet` command will determine whether there are variable names in the file, what the delimiter character is (tab or comma), and what type of data are in each column. If *filename* contains spaces, put double quotes around the filename.

If you have a text (ASCII) file that you created by saving data from a spreadsheet program, then try the `insheet` command to read that data into Stata.

By default, the `insheet` command understands files that use the tab or comma as the column delimiter. If you have a file that uses spaces as the delimiter, use the `infile` command instead. See the `infile` section of this chapter for more information. If you have a file that uses another character as the delimiter, use `insheet`'s `delimiter()` option; see [D] **insheet** for more information.

Suppose that your friend sent you the file `sample.csv`. It was saved from a spreadsheet application by saving the data as a *comma-separated values* (CSV) file. It contains the following lines:

```
VW Rabbit,4697,25,1930,3.78
Olds 98,8814,21,4060,2.41
Chev. Monza,3667,,2750,2.73
,4099,22,2930,3.58
Datsun 510,5079,24,2280,3.54
Buick Regal,5189,20,3280,2.93
Datsun 810,8129,,2750,3.55
```

These data correspond to the make, price, MPG, weight, and gear ratio of a few cars. The variable names are not in the file (so `insheet` will assign its own names), and the fields are separated by a comma. Use `insheet` to read the data:

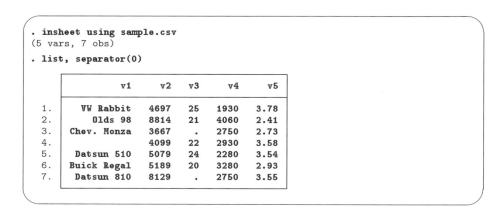

If you want to specify better variable names, you can include the desired names in the command:

```
. insheet make price mpg weight gear_ratio using sample.csv
(5 vars, 7 obs)
. list, separator(0)
```

	make	price	mpg	weight	gear_r~o
1.	VW Rabbit	4697	25	1930	3.78
2.	Olds 98	8814	21	4060	2.41
3.	Chev. Monza	3667	.	2750	2.73
4.		4099	22	2930	3.58
5.	Datsun 510	5079	24	2280	3.54
6.	Buick Regal	5189	20	3280	2.93
7.	Datsun 810	8129	.	2750	3.55

As a side note about displaying data: Stata listed gear_ratio as gear_r~o in the output from list. gear_r~o is a unique abbreviation for the variable gear_ratio. Stata displays the abbreviated variable name when variable names are longer than eight characters.

To prevent Stata from abbreviating gear_ratio, you could specify the abbreviate(10) option:

```
. list, separator(0) abbreviate(10)
```

	make	price	mpg	weight	gear_ratio
1.	VW Rabbit	4697	25	1930	3.78
2.	Olds 98	8814	21	4060	2.41
3.	Chev. Monza	3667	.	2750	2.73
4.		4099	22	2930	3.58
5.	Datsun 510	5079	24	2280	3.54
6.	Buick Regal	5189	20	3280	2.93
7.	Datsun 810	8129	.	2750	3.55

For more information on the ~ abbreviation and on list, see chapter 11.

For this simple example, you could have copied the above file and pasted it into the Data Editor, because the Data Editor also understands the comma as the delimiter.

(Continued on next page)

infile with unformatted data

The file `afewcars.raw` contains the following lines, which we can view using the `type` command:

```
. type afewcars.raw
"VW Rabbit"      4697     25      1930    3.78
"Olds 98"     8814     21      4060   2.41
"Chev. Monza"   3667     .       2750   2.73
""              4099     22      2930   3.58
"Datsun 510"    5079     24      2280   3.54
"Buick Regal"
5189
20
3280
2.93
"Datsun 810"    8129    ***      2750   3.55
```

These correspond to the make, price, MPG, weight, and gear ratio of a few cars.

This file is a mess. It uses spaces, instead of a good special character, to delimit pieces of data. The second line is not formatted the same as the first line. The sixth car's data are spread over five lines. There are some data missing, and the missing values have different representations: the MPG on the third line has a '.', the make on the fourth line contains "", and the MPG on the last line contains '***'. The one redeeming feature of the file is that the makes of cars are quoted, so that spaces in the names, such as "Olds 98", will not make the file unreadable. The `infile` command can deal with such a messy file.

```
. infile str18 make price mpg weight gear_ratio using afewcars
'***' cannot be read as a number for mpg[7]
(7 observations read)
```

Notes

1. The phrase `using afewcars` in the command is interpreted by Stata to mean `using afewcars.raw`. The extension `.raw` is assumed if you do not specify otherwise. If the data had been stored in `afewcars.txt`, you would have had to specify `using afewcars.txt`.

2. Observations need not be on one line. The location of line breaks does not matter, because they are treated the same as a space (the computer term for this is that they are treated as *whitespace*).

3. The variable names were required in this example, so that `infile` could know how many variables there were. It could not have used the number of pieces of data on the first line, because it is not treating line breaks as special.

4. Entries that are not understood (such as ***) are mentioned and are stored as missing values.

5. Entries that use Stata's conventions for missing values are not mentioned. There are two examples of this: '.', indicating a numeric missing value, and "", indicating a string missing value. However, with `infile` you must explicitly have a "" in your file so that it does not read the next number in the file as the value of this string variable.

infile with fixed-format data

File `cars2.raw` contains the following lines:

```
. type cars2.raw
VW Rabbit        4697       25      1930    3.78
Olds 98          8814       21      4060    2.41
Chev. Monza      3667               2750    2.73
                 4099       22      2930    3.58
Datsun 510       5079       24      2280    3.54
Buick Regal      5189       20      3280    2.93
Datsun 810       8129               2750    3.55
```

These data are simple for a human to read, because they are formatted nicely, but much more difficult for the computer to read without guidance. There are two related reasons for this: there are no double quotes around the strings in the first column and they include blanks, and there are blanks in the first and third columns when the data are missing. This means that `infile` cannot use whitespace as before to know where one piece of data ends and the next begins. Trying to separate the data by using whitespace leads to a complete failure, because it assigns each chunk of text to each variable in turn:

make is	VW	in the first observation	
price is	Rabbit	in the first observation	(error, stores as missing value)
MPG is	4697	in the first observation	(wrong)
weight is	25	in the first observation	(also wrong)
gear ratio is	1930	in the first observation	(wrong again)
make is	3.78	in the second observation	(stores "3.78" as a string)
price is	Olds	in the second observation	(error, stores as missing value)
MPG is	98	in the second observation	(surprise)
⋮	⋮	⋮	

This problem is referred to as "loss of synchronization".

We could solve this problem if we told `infile` that the data were nicely aligned in columns. This alignment can be passed on to `infile` in the form of a *dictionary file*. Here is a dictionary file that would work for these data. We can create it using the Do-file Editor and save it in the file `forcars2.dct`:

```
. type forcars2.dct
dictionary using cars2.raw {
        _column(1)      str18 make      %11s
        _column(19)     price           %4f
        _column(31)     mpg             %2f
        _column(39)     weight          %4f
        _column(47)     gear_ratio      %4f
}
```

You can see that the column numbers tell `infile` where to start reading each variable. Furthermore, the dictionary contains the name of the file for which it was made. The details of constructing dictionaries are described in [D] **infile (fixed format)** in the *Stata Data Management Reference Manual*. To read the data,

```
. infile using forcars2
dictionary using cars2.raw {
        _column(1)       str18 make      %11s
        _column(19)      price           %4f
        _column(31)      mpg             %2f
        _column(39)      weight          %4f
        _column(47)      gear_ratio      %4f
}
(7 observations read)
. list

                make    price   mpg   weight   gear_r~o

  1.        VW Rabbit    4697    25    1930       3.78
  2.          Olds 98    8814    21    4060       2.41
  3.      Chev. Monza    3667     .    2750       2.73
  4.                     4099    22    2930       3.58
  5.      Datsun 510     5079    24    2280       3.54

  6.      Buick Regal    5189    20    3280       2.93
  7.      Datsun 810     8129     .    2750       3.55
```

The infile command uses a dictionary because no variables are specified. It adds a file extension .dct when no extension is specified.

Stata has a second command, infix, for reading fixed-format files. Its full documentation is in the *Stata Data Management Reference Manual*.

Importing files from other software

Stata has some more specialized methods for reading data that were created or stored in another format. These are useful when you do not have control over the format in which the data are distributed.

The fdause command can read and create SAS XPORT Transport files. See fdause and fdasave in [D] **fdasave** for full details.

If you have software that supports ODBC (*Open Database Connectivity*), you can read data by using the odbc command without the need to create interim files. See [D] **odbc** for full details.

If you would like to move data using XML (*Extensible Markup Language*), Stata has the xmluse and xmlsave commands available. See xmluse and xmlsave in [D] **xmlsave** for full details.

Here is a brief series of the choices:
1. If you have a table, you could try copying it and pasting into the Data Editor.
2. If you have a file exported from a spreadsheet or database application to a tab-delimited or CSV file, use insheet.
3. If you have a free-format, space-delimited file, use infile with variable names.
4. If you have a fixed-format file, either use infile with a dictionary or use infix.
5. If you have a SAS XPORT file, use fdause.
6. If you have a database accessible via ODBC, use odbc.
7. If you have an XML file, use xmluse.
8. Finally, you can purchase a transfer program that will convert the other software's data file format to Stata's data file format. See [U] **21.4 Transfer programs**.

10 Labeling data

Making data readable

This chapter discusses, in brief, labeling of the dataset, variables, and values. Such labeling is critical to careful use of data. Labeling variables with descriptive names clarifies their meanings. Labeling values within categorical variables keeps the true meanings of the categories from becoming obscure. These points are crucial when sharing data with others, including your future self. Labels are also used in the output of most Stata commands, so proper labeling of the dataset will make much more readable results. We will work through an example of properly labeling a dataset, its variables, and the values of one encoded variable.

The dataset structure: describe

In chapter 9, we saved a dataset called `afewcars.dta`. We will get this dataset into a shape that a colleague would understand. Let's see what it contains.

```
. use afewcars
. list, separator(0)

            make    price    mpg    weight    gear_r~o

  1.    VW Rabbit    4697     25      1930        3.78
  2.      Olds 98    8814     21      4060        2.41
  3.  Chev. Monza    3667      .      2750        2.73
  4.                 4099     22      2930        3.58
  5.    Datsun 510    5079     24      2280        3.54
  6.   Buick Regal    5189     20      3280        2.93
  7.    Datsun 810    8129      .      2750        3.55
```

The data allow us to make some guesses at the values in the dataset, but, for example, we do not know the units in which the price or weight is measured, and the term "mpg" could be confusing for people outside the United States. Perhaps we can learn something from the description of the dataset. Stata has the aptly named `describe` command for this purpose (as we saw in chap. 3).

(Continued on next page)

```
. describe

Contains data from afewcars.dta
  obs:            7
  vars:           5                                    13 Apr 2007 15:19
  size:         266  (99.9% of memory free)
─────────────────────────────────────────────────────────────────────────
              storage   display     value
variable name   type    format      label      variable label
─────────────────────────────────────────────────────────────────────────
make          str18    %18s
price         float    %9.0g
mpg           float    %9.0g
weight        float    %9.0g
gear_ratio    float    %9.0g
─────────────────────────────────────────────────────────────────────────
Sorted by:
```

Though there is precious little information that could help us as a researcher, we can glean some information here about how Stata thinks of the data from the first three columns of the output.

1. The *variable name* is the name we use to tell Stata about a variable.

2. The *storage type* (otherwise known as the *data type*) is the way in which Stata stores the data in a variable. There are six different storage types:

   a. For integers:

      byte for integers between −127 and 100

      int for integers between −32,767 and 32,740

      long for integers between −2,147,483,647 and 2,147,483,620

   b. For real numbers:

      float for real numbers with 8.5 digits of precision

      double for real numbers with 16.5 digits of precision

   c. For strings (text) from 1–244 characters:

      str1 for one-character-long strings

      str2 for two-character-long strings

      str3 for three-character-long strings

      . . .

      str244 for 244-character-long strings

Storage types affect both the precision of computations and the size of datasets. A quick guide to storage types is available at help datatype; complete coverage is given in [D] **data types**.

3. The *display format* controls how the variable is displayed; see [U] **12.5 Formats: controlling how data are displayed**. By default, Stata sets it to something reasonable given the storage type. We would like to make this dataset into something containing all the information we need.

To see what a well-labeled dataset looks like, we can take a look at a dataset stored at the Stata Press repository. We need not load the data (and disturb what we are doing); we do not even need a copy of the dataset on our machine. (You will learn more about Stata's Internet capabilities in chap. 20.) All we need to do is to direct describe to look at the proper file by using the command describe using *filename*.

```
. describe using http://www.stata-press.com/data/r10/auto
Contains data                               1978 Automobile Data
  obs:            74                         13 Apr 2007 17:45
  vars:           12
  size:        3,478

              storage   display    value
variable name   type    format     label     variable label

make           str18    %-18s                Make and Model
price          int      %8.0gc               Price
mpg            int      %8.0g                Mileage (mpg)
rep78          int      %8.0g                Repair Record 1978
headroom       float    %6.1f                Headroom (in.)
trunk          int      %8.0g                Trunk space (cu. ft.)
weight         int      %8.0gc               Weight (lbs.)
length         int      %8.0g                Length (in.)
turn           int      %8.0g                Turn Circle (ft.)
displacement   int      %8.0g                Displacement (cu. in.)
gear_ratio     float    %6.2f                Gear Ratio
foreign        byte     %8.0g      origin    Car type

Sorted by:  foreign
```

This output is much more informative. There are three locations where labels are attached that help explain what the dataset contains:

1. In the first line, 1978 Automobile Data is the *data label.* It gives information about the contents of the dataset. Data can be labeled by selecting **Data > Labels > Label dataset** or by using the `label data` command.

2. There is a *variable label* attached to each variable. Variable labels are not only how we would refer to the variable in normal everyday conversation, but they also contain information about the units of the variables. Variables can be labeled by selecting **Data > Labels > Label variable** or by using the `label variable` command.

3. The `foreign` variable has an attached *value label.* Value labels allow numeric variables, such as `foreign`, to have words associated with numeric codes. The `describe` tells you that the numeric variable `foreign` has value label `origin` associated with it. Although not revealed by `describe`, the variable `foreign` takes on the values 0 and 1, and the value label `origin` associates 0 with Domestic and 1 with Foreign. When you `browse` the data (see chap. 8), `foreign` appears to contain the values "Domestic" and "Foreign". It is not necessary for the value label to have a name different from that of the variable. You could just as easily have used a value label named `foreign`. The values in a variable are labeled in two stages. The value label must first be defined. This can be done by selecting **Data > Labels > Label values > Define or modify value labels** or by typing the `label define` command. After the labels have been defined, they must be attached to the proper variables either by selecting **Data > Labels > Label values > Assign value labels to variable** or by using the `label values` command.

(Continued on next page)

Labeling datasets and variables

We will now take our `afewcars.dta` dataset and give it proper labels. We will do this via the command window, because it is simple to do in this fashion. If you use the menus and dialogs, you will end up with the same commands in your log.

```
. use afewcars
. describe
Contains data from afewcars.dta
  obs:           7
  vars:          5                                13 Apr 2007 15:19
  size:        266 (99.9% of memory free)

              storage  display    value
variable name  type    format     label     variable label

make          str18   %18s
price         float   %9.0g
mpg           float   %9.0g
weight        float   %9.0g
gear_ratio    float   %9.0g

Sorted by:
. label data "A few 1978 cars"
. label variable make "Make and Model"
. label variable price "Price (USD)"
. label variable mpg "Mileage (mile per gallon)"
. label variable weight "Vehicle weight (lbs.)"
. label variable gear_ratio "Gear Ratio"
. describe
Contains data from afewcars.dta
  obs:           7                                A few 1978 cars
  vars:          5                                13 Apr 2007 15:19
  size:        266 (99.9% of memory free)

              storage  display    value
variable name  type    format     label     variable label

make          str18   %18s                  Make and Model
price         float   %9.0g                 Price (USD)
mpg           float   %9.0g                 Mileage (mile per gallon)
weight        float   %9.0g                 Vehicle weight (lbs.)
gear_ratio    float   %9.0g                 Gear Ratio

Sorted by:
. save, replace
file afewcars.dta saved
```

Warning: when you change or define labels on a dataset in memory, it is worth saving the dataset right away. Since the actual data in the dataset did not change, Stata will not prevent you from exiting or loading a new dataset later, and you will lose your labels.

Labeling values of variables

Suppose that we now added a new indicator variable to the dataset that is 0 if the car was made in the United States and 1 if it was foreign. Our data would now look like this:

```
. list, separator(0)

             make    price    mpg   weight   gear_r~o   foreign

  1.     VW Rabbit    4697     25     1930      3.78         1
  2.       Olds 98    8814     21     4060      2.41         0
  3.   Chev. Monza    3667      .     2750      2.73         0
  4.                  4099     22     2930      3.58         0
  5.    Datsun 510    5079     24     2280      3.54         1
  6.   Buick Regal    5189     20     3280      2.93         0
  7.    Datsun 810    8129      .     2750      3.55         1
```

Though the definition of the categories "0" and "1" are clear in this context, it still would be worthwhile to give the values explicit labels, because it will make output clear to people who are not so familiar with indicator variables. This is done with a value label.

```
. label define origin 0 "domestic" 1 "foreign"
. label values foreign origin
. describe
Contains data from afewcars.dta
  obs:              7                      A few 1978 cars
  vars:             6                      13 Apr 2007 15:19
  size:           273  (99.9% of memory free)

               storage   display    value
variable name   type     format     label      variable label

make            str18    %18s                   Make and Model
price           float    %9.0g                  Price (USD)
mpg             float    %9.0g                  Mileage (mile per gallon)
weight          float    %9.0g                  Vehicle weight (lbs.)
gear_ratio      float    %9.0g                  Gear Ratio
foreign         byte     %8.0g      origin

Sorted by:
     Note:  dataset has changed since last saved
. save, replace
file afewcars.dta saved
```

From this we can see that a value label is defined via

    label define *labelname* # "*contents*" # "*contents*" ...

It can then be attached to a variable via

    label values *variablename labelname*

Once again, we need to save the dataset, to be sure that we do not mistakenly lose the labels later.

Labels using menus and dialogs

You can open the **Data > Labels** menu for a list of dialogs that can add, delete, define, or modify labels. For example, to label a dataset, select **Data > Labels > Label dataset** to use the *Label dataset* dialog. To label a variable, select **Data > Labels > Label variable** to use the *Label variable* dialog.

For some commands, the dialogs are not any more efficient than typing the command by hand, but they may be helpful when you can't remember the syntax of the command. For example, instead of typing `label define` *labelname # "contents" # "contents"* ... to create a new value label, you could select **Data > Labels > Label values > Define or modify value labels** to open the *Define value labels* dialog. From this dialog,

1. Click the **Define...** button.

2. Type a label name, and click **OK**.

3. Type a numeric value in the *Value* edit field from the *Add value* dialog.

4. Type the value label in the *Text* edit field, and click **OK**.

5. Repeat steps 3 and 4.

6. Click **Cancel** when you are finished adding values.

You can always open the *Define value labels* dialog again to add or modify more value labels.

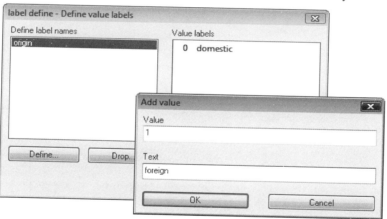

Notes

There is more to value labels than what was covered here. See [U] **12.6.3 Value labels** for a complete treatment.

You may also add notes to your data and your variables. This feature was not discussed here, but you can learn about notes by typing `help notes` or get the full story in [D] **notes**.

11 Listing data and basic command syntax

Command syntax

This chapter gives a basic lesson on Stata's command syntax while showing how to control the appearance of a data list.

As we've seen throughout this book, the user has a choice between using menus and dialogs and using the Command window. Although many find the menus more natural and the Command window baffling at first, some practice makes working via the Command window often much faster than using menus and dialogs. The Command window can become a faster way of working because of the clean and regular syntax of Stata commands. We will cover enough to get you started; `help language` has more information and examples, and [U] **11 Language syntax** has all the details.

The syntax for the `list` command can be seen by typing `help list`:

$$\underline{\text{l}}\text{ist} \; \big[\textit{varlist}\big] \; \big[\textit{if}\big] \; \big[\textit{in}\big] \; \big[\, ,\, \textit{options}\big]$$

The way that this can be read is the following:
- Anything inside square brackets is optional. For the `list` command,
  — *varlist* is optional. A *varlist* is a list of variable names.
  — *if* is optional. The `if` qualifier restricts the command to run only on those observations for which the qualifier is true. We saw examples of this in chapter 8.
  — *in* is optional. The `in` qualifier restricts the command to run on particular observation numbers.
  — , *options* are optional and are separated from the rest of the command by a comma.
- Optional pieces do not preclude one another unless explicitly stated. For the `list` command, it is possible to use a *varlist* with *if* and *in*.
- If a part of a word is underlined, the underlined part is the minimum abbreviation. Any abbreviation at least this long is acceptable.
  — The l in `list` is underlined, so l, li, and lis are all equivalent to `list`.
- Anything not inside square brackets is required. For the `list` command, only the command itself is required.

Keeping these rules in mind, let's investigate how `list` behaves when called with different arguments. We will be using the dataset `afewcars.dta` from the end of last chapter.

list with a variable list

Variable lists (or *varlists*) can be specified in a variety of ways, all designed to save typing and encourage good variable names.
- The *varlist* is optional for `list`. This means that if no variables are specified, it is equivalent to specifying all variables. Another way to think of it is that the default behavior of the command is to run on all variables, unless restricted by a *varlist*.
- You can list a subset of variables explicitly, as in `list make mpg price`.
- There are many shorthand notations, also:
  `m*` means all variables starting with m.
  `price-weight` means all variables from `price` through `weight` in dataset order.
  `ma?e` means all variables starting with `ma`, followed by any character, and ending in `e`.

- You can list a variable by using an abbreviation unique to that variable, as in list gear_r~o. If the abbreviation is not unique, Stata returns an error message.

```
. list

              make    price   mpg   weight   gear_r~o   foreign

  1.    VW Rabbit     4697     25     1930       3.78    foreign
  2.       Olds 98    8814     21     4060       2.41    domestic
  3.   Chev. Monza    3667      .     2750       2.73    domestic
  4.                  4099     22     2930       3.58    domestic
  5.    Datsun 510    5079     24     2280       3.54    foreign

  6.   Buick Regal    5189     20     3280       2.93    domestic
  7.    Datsun 810    8129      .     2750       3.55    foreign

. l make mpg price

              make    mpg   price

  1.    VW Rabbit      25    4697
  2.       Olds 98     21    8814
  3.   Chev. Monza      .    3667
  4.                   22    4099
  5.    Datsun 510     24    5079

  6.   Buick Regal     20    5189
  7.    Datsun 810      .    8129

. list m*

              make    mpg

  1.    VW Rabbit      25
  2.       Olds 98     21
  3.   Chev. Monza      .
  4.                   22
  5.    Datsun 510     24

  6.   Buick Regal     20
  7.    Datsun 810      .

. li price-weight

      price   mpg   weight

  1.   4697    25     1930
  2.   8814    21     4060
  3.   3667     .     2750
  4.   4099    22     2930
  5.   5079    24     2280

  6.   5189    20     3280
  7.   8129     .     2750
```

```
. list ma?e

            make

1.     VW Rabbit
2.       Olds 98
3.   Chev. Monza
4.
5.     Datsun 510

6.    Buick Regal
7.    Datsun 810


. l gear_r~o

         gear_r~o

1.          3.78
2.          2.41
3.          2.73
4.          3.58
5.          3.54

6.          2.93
7.          3.55
```

list with if

The if qualifier uses a logical expression to determine which observations to use. If the expression is true, the observation is used in the command; otherwise, it is skipped. The operators whose results are either true or false are

<	less than
<=	less than or equal
==	equal
>	greater than
>=	greater than or equal
!=	not equal (~= can also be used)
&	and
|	or
!	not (logical negation; ~ can also be used)
()	parentheses are for grouping to specify order of evaluation

In the logical expressions, & is evaluated before | (similar to multiplication before addition in arithmetic). You can use this in your expressions, but it is often better to use parentheses to ensure that the expressions are evaluated properly. See [U] **13.2 Operators** for the complete story.

```
. list

          make   price   mpg   weight   gear_r~o   foreign

  1.    VW Rabbit   4697    25     1930       3.78   foreign
  2.      Olds 98   8814    21     4060       2.41   domestic
  3.   Chev. Monza   3667     .     2750       2.73   domestic
  4.                 4099    22     2930       3.58   domestic
  5.    Datsun 510   5079    24     2280       3.54   foreign

  6.   Buick Regal   5189    20     3280       2.93   domestic
  7.    Datsun 810   8129     .     2750       3.55   foreign

. list if mpg > 22

          make   price   mpg   weight   gear_r~o   foreign

  1.    VW Rabbit   4697    25     1930       3.78   foreign
  3.   Chev. Monza   3667     .     2750       2.73   domestic
  5.    Datsun 510   5079    24     2280       3.54   foreign
  7.    Datsun 810   8129     .     2750       3.55   foreign

. list if (mpg > 22) & !missing(mpg)

          make   price   mpg   weight   gear_r~o   foreign

  1.    VW Rabbit   4697    25     1930       3.78   foreign
  5.    Datsun 510   5079    24     2280       3.54   foreign

. list make mpg price gear if (mpg > 22) | (price > 8000 & gear < 3.5)

          make   mpg   price   gear_r~o

  1.    VW Rabbit    25    4697       3.78
  2.      Olds 98    21    8814       2.41
  3.   Chev. Monza     .    3667       2.73
  5.    Datsun 510    24    5079       3.54
  7.    Datsun 810     .    8129       3.55

. list make mpg if mpg <= 22 in 2/4

        make   mpg

  2.   Olds 98    21
  4.              22
```

In the listings above, we see more examples of Stata treating missing values as large values, as well as the care that should be taken when the if qualifier is applied to a variable with missing values. See chapter 8.

list with if, common mistakes

Here is a series of listings with common errors and their corrections. See if you can find the errors before reading the correct entry.

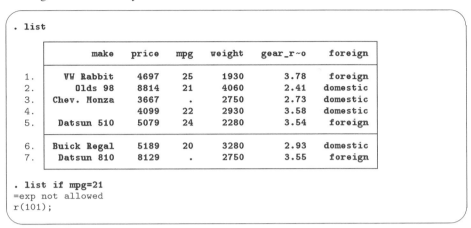

```
. list

              make   price   mpg   weight   gear_r~o   foreign

  1.      VW Rabbit    4697    25     1930       3.78    foreign
  2.        Olds 98    8814    21     4060       2.41   domestic
  3.    Chev. Monza    3667     .     2750       2.73   domestic
  4.                   4099    22     2930       3.58   domestic
  5.     Datsun 510    5079    24     2280       3.54    foreign

  6.    Buick Regal    5189    20     3280       2.93   domestic
  7.     Datsun 810    8129     .     2750       3.55    foreign

. list if mpg=21
=exp not allowed
r(101);
```

The error arises because "equal" is expressed by ==, not by =. Corrected, it becomes

```
. list if mpg==21

           make   price   mpg   weight   gear_r~o   foreign

  2.     Olds 98    8814    21     4060       2.41   domestic
```

Another common error with logic:

```
. list if mpg==21 if weight > 4000
invalid syntax
r(198);
. list if mpg==21 and weight > 4000
invalid 'and'
r(198);
```

Joint tests are specified with &, not with the word and or multiple ifs. The if qualifier should be if mpg==21 & weight>4000, not if mpg==21 if weight>4000. Here is its correction:

```
. list if mpg==21 & weight > 4000

           make   price   mpg   weight   gear_r~o   foreign

  2.     Olds 98    8814    21     4060       2.41   domestic
```

A problem with string variables:

```
. list if make==Datsun 510
Datsun not found
r(111);
```

Strings must be in double quotes, as in make=="Datsun 510". Without the quotes, Stata thinks that Datsun is a variable that it cannot find. Here is the correction:

```
. list if make=="Datsun 510"

          make    price   mpg   weight   gear_r~o   foreign

   5.   Datsun 510   5079    24     2280       3.54   foreign
```

Confusing value labels with strings:

```
. list if foreign=="domestic"
type mismatch
r(109);
```

Value labels look like strings but the underlying variable is numeric. Variable foreign takes on values 0 and 1 but has the value label that attaches 0 to "domestic" and 1 to "foreign" (see chap. 10). To see the underlying numeric values of variables with labeled values, use the label list command (see [D] **label**), or investigate the variable with codebook *varname*. We can correct the error here by looking for observations where foreign==0.

There is a second construction that also allows the use of the value label directly.

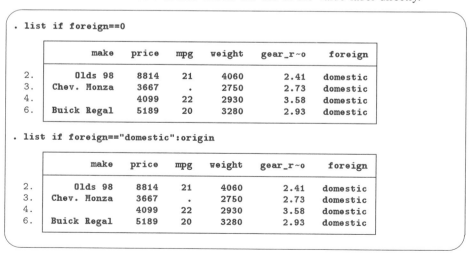

```
. list if foreign==0

          make    price   mpg   weight   gear_r~o   foreign

   2.      Olds 98    8814    21     4060       2.41   domestic
   3.   Chev. Monza   3667     .     2750       2.73   domestic
   4.                 4099    22     2930       3.58   domestic
   6.   Buick Regal   5189    20     3280       2.93   domestic

. list if foreign=="domestic":origin

          make    price   mpg   weight   gear_r~o   foreign

   2.      Olds 98    8814    21     4060       2.41   domestic
   3.   Chev. Monza   3667     .     2750       2.73   domestic
   4.                 4099    22     2930       3.58   domestic
   6.   Buick Regal   5189    20     3280       2.93   domestic
```

list with in

The `in` qualifier uses a *numlist* to give a range of observations that should be listed. *numlists* have the form of one number or *first/last*. Positive numbers count from the top of the dataset. Negative numbers count from the end of the dataset. Here are some examples:

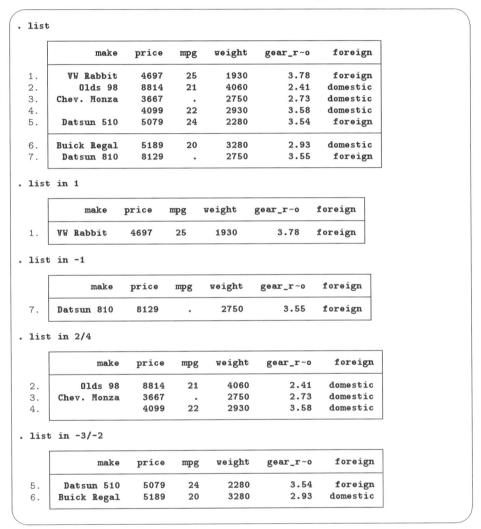

Controlling the list output

The fine control over `list` output is exercised by specifying one or more options. You can use `sepby()` to separate observations by variable. `abbreviate()` specifies the minimum number of characters to abbreviate a variable in the output. `divider` draws a vertical line between the variables in the list.

```
. sort foreign
. list ma p g f, sepby(foreign)

             make      price    gear_r~o     foreign

  1.    Chev. Monza     3667        2.73     domestic
  2.                    4099        3.58     domestic
  3.    Buick Regal     5189        2.93     domestic
  4.         Olds 98    8814        2.41     domestic

  5.    Datsun 510      5079        3.54      foreign
  6.     VW Rabbit      4697        3.78      foreign
  7.    Datsun 810      8129        3.55      foreign

. list make weight gear, abbreviate(10)

             make     weight    gear_ratio

  1.    Chev. Monza     2750        2.73
  2.                    2930        3.58
  3.    Buick Regal     3280        2.93
  4.         Olds 98    4060        2.41
  5.    Datsun 510      2280        3.54

  6.     VW Rabbit      1930        3.78
  7.    Datsun 810      2750        3.55

. list, divider

             make  |  price  |  mpg  |  weight  |  gear_r~o  |  foreign

  1.    Chev. Monza    3667      .       2750        2.73      domestic
  2.                   4099     22       2930        3.58      domestic
  3.    Buick Regal    5189     20       3280        2.93      domestic
  4.         Olds 98   8814     21       4060        2.41      domestic
  5.    Datsun 510     5079     24       2280        3.54       foreign

  6.     VW Rabbit     4697     25       1930        3.78       foreign
  7.    Datsun 810     8129      .       2750        3.55       foreign
```

The separator() option draws a horizontal line at specified intervals. When not specified, it defaults to a value of 5.

```
. list, separator(3)

               make   price    mpg   weight   gear_r~o    foreign

  1.    Chev. Monza     3667      .     2750       2.73    domestic
  2.                    4099     22     2930       3.58    domestic
  3.    Buick Regal     5189     20     3280       2.93    domestic

  4.        Olds 98     8814     21     4060       2.41    domestic
  5.     Datsun 510     5079     24     2280       3.54     foreign
  6.      VW Rabbit     4697     25     1930       3.78     foreign

  7.     Datsun 810     8129      .     2750       3.55     foreign
```

More

When you see a —more— at the bottom of the Results window, it means that there is more information to be displayed. This happens, for example, when you are listing many observations.

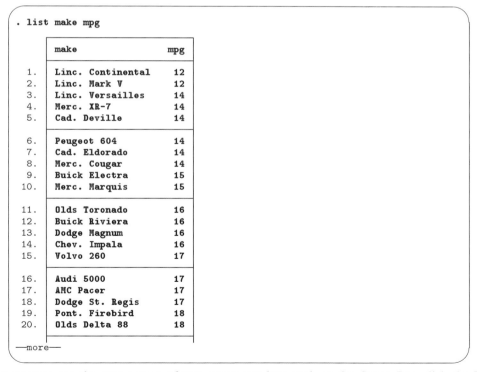

```
. list make mpg

               make         mpg

  1.    Linc. Continental    12
  2.    Linc. Mark V         12
  3.    Linc. Versailles     14
  4.    Merc. XR-7           14
  5.    Cad. Deville         14

  6.    Peugeot 604          14
  7.    Cad. Eldorado        14
  8.    Merc. Cougar         14
  9.    Buick Electra        15
 10.    Merc. Marquis        15

 11.    Olds Toronado        16
 12.    Buick Riviera        16
 13.    Dodge Magnum         16
 14.    Chev. Impala         16
 15.    Volvo 260            17

 16.    Audi 5000            17
 17.    AMC Pacer            17
 18.    Dodge St. Regis      17
 19.    Pont. Firebird       18
 20.    Olds Delta 88        18

—more—
```

If you want to see the next screen of text, press any key, such as the *Space Bar*, click the **More** button, 【GO】, or click on the blue —more— at the bottom of the Results window. To see just the next line of text, press *Enter*.

Break

If you want to interrupt a Stata command, click the **Break** button, 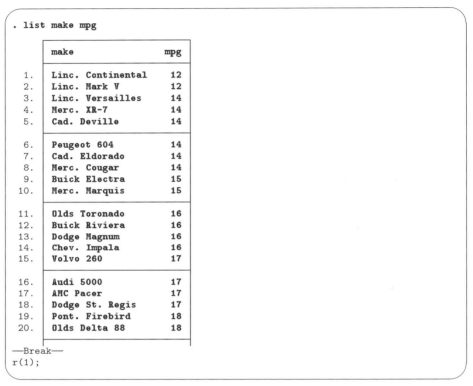. If you see a —more— at the bottom of the Results window and wish to interrupt it, click the **Break** button or press q.

```
. list make mpg

        +----------------------------+
        | make                  mpg  |
        |----------------------------|
     1. | Linc. Continental      12  |
     2. | Linc. Mark V           12  |
     3. | Linc. Versailles       14  |
     4. | Merc. XR-7             14  |
     5. | Cad. Deville           14  |
        |----------------------------|
     6. | Peugeot 604            14  |
     7. | Cad. Eldorado          14  |
     8. | Merc. Cougar           14  |
     9. | Buick Electra          15  |
    10. | Merc. Marquis          15  |
        |----------------------------|
    11. | Olds Toronado          16  |
    12. | Buick Riviera          16  |
    13. | Dodge Magnum           16  |
    14. | Chev. Impala           16  |
    15. | Volvo 260              17  |
        |----------------------------|
    16. | Audi 5000              17  |
    17. | AMC Pacer              17  |
    18. | Dodge St. Regis        17  |
    19. | Pont. Firebird         18  |
    20. | Olds Delta 88          18  |
        +----------------------------+
—Break—
r(1);
```

It is always safe to click **Break**. After you click **Break**, the state of the system is the same as if you had never issued the original command.

12 Creating new variables

generate and replace

This chapter shows the basics of creating and modifying variables in Stata. The two primary commands used for this are

- `generate` for creating new variables. It has a minimum abbreviation of g.
- `replace` for replacing the values of an existing variable. It may not be abbreviated, because it alters existing data, and hence can be considered dangerous.

The most basic form for creating new variables is `generate` *newvar* = *exp*, where *exp* is any kind of *expression*, but both can be used with `if` and `in` qualifiers, of course. An expression is a formula made up of constants, existing variables, operators, and functions. Some examples of expressions (using variables from the `auto` dataset) would be 2 + price, weight^2, or sqrt(gear_ratio).

The operators defined in Stata are given in the table below:

	Arithmetic		Logical		Relational (numeric and string)
+	addition	!	not	>	greater than
−	subtraction	|	or	<	less than
*	multiplication	&	and	>=	> or equal
/	division			<=	< or equal
^	power			==	equal
				!=	not equal
				~=	not equal
+	string concatenation				

Stata has many mathematical, statistical, string, date, time-series, and programming functions. See `help functions` for the basics, and see [D] **functions** for a complete list and full details of all the built-in functions.

You can use menus and dialog boxes to create new variables and modify existing variables by selecting menu items from the **Data > Create or change variables** menu. This feature can be handy for finding functions quickly. We will use the Command window for the examples in this chapter, however, because we would like to illustrate simple usage and some pitfalls.

generate

There are some details you should know about the `generate` command:

1. The basic form of the `generate` command is `generate` *newvar* = *exp*, where *newvar* is a new variable name, and *exp* is any valid expression. You will get an error message if you try to `generate` a variable that already exists.
2. An algebraic calculation using a missing value yields a missing value, as does division by zero, etc.

3. If missing values are generated, the number of missing values in *newvar* is always reported. If Stata says nothing about missing values, no missing values were generated.

4. You can use `generate` to set the storage type of the new variable as it is generated. Saving an indicator variable as a `byte` can save size when used with a large dataset.

Here are some examples of creating new variables from the `afewcars` dataset we've been using for the past few chapters. Some are nonsensical—they are only for illustrating the `generate` command. Why is it important in the last example to use the `if` qualifier?

```
. list make mpg weight

              make    mpg    weight

  1.      VW Rabbit     25      1930
  2.         Olds 98    21      4060
  3.    Chev. Monza      .      2750
  4.                    22      2930
  5.     Datsun 510     24      2280

  6.    Buick Regal     20      3280
  7.     Datsun 810      .      2750

. generate lphk = 3.7854 * (100 / 1.6093) / mpg
(2 missing values generated)
. label var lphk "Liters per 100km"
. g strange = sqrt(mpg*weight)
(2 missing values generated)
. gen huge = weight >= 3000 if !missing(weight)
. l make mpg weight lphk strange huge

              make    mpg    weight        lphk      strange    huge

  1.      VW Rabbit     25      1930    9.408812     219.6588       0
  2.         Olds 98    21      4060    11.20097     291.9932       1
  3.    Chev. Monza      .      2750           .            .       0
  4.                    22      2930    10.69183     253.8897       0
  5.     Datsun 510     24      2280    9.800845     233.9231       0

  6.    Buick Regal     20      3280    11.76101      256.125       1
  7.     Datsun 810      .      2750           .            .       0
```

replace

Whereas `generate` is used to create new variables, `replace` is the command used for existing variables. Stata uses two different commands to prevent you from accidentally modifying your data. The `replace` command cannot be abbreviated. Stata generally requires you to spell out completely any command that can alter your existing data.

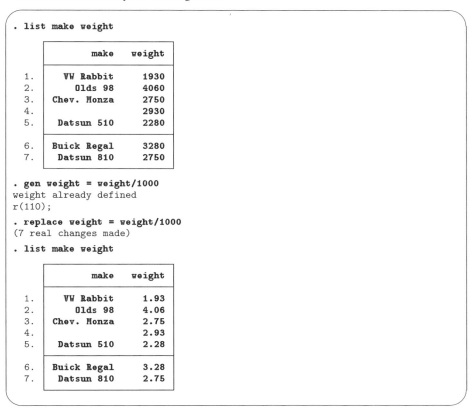

```
. list make weight

            make    weight

1.      VW Rabbit      1930
2.        Olds 98      4060
3.    Chev. Monza      2750
4.                     2930
5.     Datsun 510      2280

6.    Buick Regal      3280
7.     Datsun 810      2750

. gen weight = weight/1000
weight already defined
r(110);
. replace weight = weight/1000
(7 real changes made)
. list make weight

            make    weight

1.      VW Rabbit      1.93
2.        Olds 98      4.06
3.    Chev. Monza      2.75
4.                     2.93
5.     Datsun 510      2.28

6.    Buick Regal      3.28
7.     Datsun 810      2.75
```

Suppose that you want to create a new variable, `predprice`, which will be the predicted price of the cars in the following year. You estimate that domestic cars will increase in price by 5% and foreign cars by 10%.

One way to create the variable would be to first use `generate` to compute the predicted domestic car prices. Then use `replace` to change the missing values for the foreign cars to their proper values.

(Continued on next page)

```
. gen predprice = 1.05*price if foreign==0
(3 missing values generated)
. replace predprice = 1.10*price if foreign==1
(3 real changes made)
. list make foreign price predprice, nolabel
```

	make	foreign	price	predpr~e
1.	VW Rabbit	1	4697	5166.7
2.	Olds 98	0	8814	9254.7
3.	Chev. Monza	0	3667	3850.35
4.		0	4099	4303.95
5.	Datsun 510	1	5079	5586.9
6.	Buick Regal	0	5189	5448.45
7.	Datsun 810	1	8129	8941.9

Of course, since foreign is an indicator variable, we could generate the predicted variable with one command:

```
. gen predprice2 = (1.05 + 0.05*foreign)*price
. list make foreign price predprice predprice2, nolabel
```

	make	foreign	price	predpr~e	predpr~2
1.	VW Rabbit	1	4697	5166.7	5166.7
2.	Olds 98	0	8814	9254.7	9254.7
3.	Chev. Monza	0	3667	3850.35	3850.35
4.		0	4099	4303.95	4303.95
5.	Datsun 510	1	5079	5586.9	5586.9
6.	Buick Regal	0	5189	5448.45	5448.45
7.	Datsun 810	1	8129	8941.9	8941.9

generate with string variables

Stata is smart. When you generate a variable and the expression evaluates to a string, Stata creates a string variable with a storage type as long as necessary, and no longer than that. `where` is a `str1` in the following example.

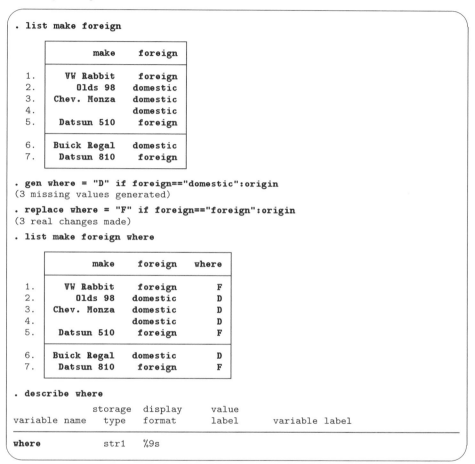

```
. list make foreign

              make    foreign

  1.      VW Rabbit    foreign
  2.         Olds 98   domestic
  3.    Chev. Monza    domestic
  4.                   domestic
  5.     Datsun 510    foreign

  6.    Buick Regal    domestic
  7.     Datsun 810    foreign

. gen where = "D" if foreign=="domestic":origin
(3 missing values generated)
. replace where = "F" if foreign=="foreign":origin
(3 real changes made)
. list make foreign where

              make    foreign   where

  1.      VW Rabbit    foreign     F
  2.         Olds 98   domestic    D
  3.    Chev. Monza    domestic    D
  4.                   domestic    D
  5.     Datsun 510    foreign     F

  6.    Buick Regal    domestic    D
  7.     Datsun 810    foreign     F

. describe where

                storage   display     value
variable name    type     format      label        variable label

where            str1     %9s
```

Stata has some useful tools for working with string variables. Here we split the `make` variable into make and model and then create a variable that has the model together with where the model was manufactured.

(Continued on next page)

```
. gen model = substr(make, strpos(make," ")+1, .)
(1 missing value generated)

. gen modelwhere = model + " " + where

. list make where model modelwhere
```

	make	where	model	modelw~e
1.	VW Rabbit	F	Rabbit	Rabbit F
2.	Olds 98	D	98	98 D
3.	Chev. Monza	D	Monza	Monza D
4.		D		D
5.	Datsun 510	F	510	510 F
6.	Buick Regal	D	Regal	Regal D
7.	Datsun 810	F	810	810 F

There are a few things to note about how these commands worked:

1. $strpos(s_1,s_2)$ produces an integer equal to the first position in s_1 at which s_2 is found, or 0 if it is not found.

2. $substr(s,c_0,c_1)$ produces a string of length c_1 beginning with the c_0'th character of s. If $c_1 = .$, it gives a string from character c_0 to the end of string.

3. Putting 1 and 2 together: $substr(s,strpos(s," ")+1,.)$ produces s with its first word removed. Since make contains both the make and model of each car, and make contains no spaces in this dataset, we have found each car's model.

4. The operator '+' when applied to string variables will concatenate the strings (i.e., join them together). The expression "this" + "that" results in the string "thisthat". When the variable modelwhere was generated, a space (" ") was added between the two strings.

5. The missing value for a string is nothing special—it is simply the empty string "". Thus, the value of modelwhere for the car with no make or model is " D" (note the leading space).

13 Deleting variables and observations

clear, drop, and keep

In this chapter we will present the tools for paring observations and variables from a dataset. We saw how to do this using the Data Editor in chapter 8; this chapter presents the methods for doing so from the Command window.

There are three main commands for removing data and other Stata objects from memory: `clear`, `drop`, and `keep`. Remember that they affect only what is in memory. None of these commands alters anything that has been saved to disk.

clear and drop _all

Suppose that you are working on an analysis or a simulation, and you need to clear out Stata's memory, so that you can impute different values or simulate a new dataset. You are not interested in saving any of the changes you have made to the dataset in memory—you would just like to have an empty dataset. What you do depends on how much you want to clear out: at any time, you can have not only data but also metadata such as value labels, saved results from previous commands, and saved matrices. The `clear` command will let you carefully clear out data or other objects; we are interested only in simple usage here. For more information, see `help clear` and [D] **clear**.

If you type the command `clear` into the Command window, it will remove all variables and value labels. In basic usage, this is typically enough. It has the nice property that it does not remove any saved results, so you can load a new dataset and predict values by using saved estimation results from a model fitted on a previous dataset. See `help postest` and [U] **20 Estimation and postestimation commands** for more information.

If you want to be sure that everything is cleared out, use the command `clear all`. This will clear Stata's memory of data and all auxiliary objects, so that you can start with a clean slate. The first time you use `clear all` while you have a graph or dialog open, you may be surprised when that graph or dialog closes; this is necessary so that Stata can free all memory that is being used.

If you want to get rid of just the data and nothing else, you can use the command `drop _all`.

drop

The `drop` command is used to remove variables or observations from the dataset in memory.
- If you want to drop variables, use `drop` *varlist*.
- If you want to drop observations, use `drop` with an `if` and/or an `in` qualifier.

We will use the afewcars dataset, as usual.

```
. use afewcars
(A few 1978 cars)
. list
```

	make	price	mpg	weight	gear_r~o	foreign
1.	VW Rabbit	4697	25	1930	3.78	foreign
2.	Olds 98	8814	21	4060	2.41	domestic
3.	Chev. Monza	3667	.	2750	2.73	domestic
4.		4099	22	2930	3.58	domestic
5.	Datsun 510	5079	24	2280	3.54	foreign
6.	Buick Regal	5189	20	3280	2.93	domestic
7.	Datsun 810	8129	.	2750	3.55	foreign

```
. drop in 1/3
(3 observations deleted)
. list
```

	make	price	mpg	weight	gear_r~o	foreign
1.		4099	22	2930	3.58	domestic
2.	Datsun 510	5079	24	2280	3.54	foreign
3.	Buick Regal	5189	20	3280	2.93	domestic
4.	Datsun 810	8129	.	2750	3.55	foreign

```
. drop if mpg > 21
(3 observations deleted)
. list
```

	make	price	mpg	weight	gear_r~o	foreign
1.	Buick Regal	5189	20	3280	2.93	domestic

```
. drop gear_ratio
. list
```

	make	price	mpg	weight	foreign
1.	Buick Regal	5189	20	3280	domestic

```
. drop m*
. list
```

	price	weight	foreign
1.	5189	3280	domestic

These changes are only to the data in memory. If you wanted to make the changes permanent, you would need to save the dataset.

keep

keep tells Stata to drop all variables except those specified explicitly or through the use of an `if` or `in` expression. Just like `drop`, it can be used with a *varlist* or with qualifiers, but not with both at once. We use a `clear` command at the start of this example, so that we can reload the `afewcars` dataset.

```
. clear
. use afewcars
(A few 1978 cars)
. list
```

	make	price	mpg	weight	gear_r~o	foreign
1.	VW Rabbit	4697	25	1930	3.78	foreign
2.	Olds 98	8814	21	4060	2.41	domestic
3.	Chev. Monza	3667	.	2750	2.73	domestic
4.		4099	22	2930	3.58	domestic
5.	Datsun 510	5079	24	2280	3.54	foreign
6.	Buick Regal	5189	20	3280	2.93	domestic
7.	Datsun 810	8129	.	2750	3.55	foreign

```
. keep in 4/7
(3 observations deleted)
. list
```

	make	price	mpg	weight	gear_r~o	foreign
1.		4099	22	2930	3.58	domestic
2.	Datsun 510	5079	24	2280	3.54	foreign
3.	Buick Regal	5189	20	3280	2.93	domestic
4.	Datsun 810	8129	.	2750	3.55	foreign

```
. keep if mpg <= 21
(3 observations deleted)
. list
```

	make	price	mpg	weight	gear_r~o	foreign
1.	Buick Regal	5189	20	3280	2.93	domestic

```
. keep m*
. list
```

	make	mpg
1.	Buick Regal	20

Notes

14 Using the Do-file Editor—automating Stata

The Do-file Editor

Stata comes with an integrated text editor called the **Do-file Editor**, which can be used for many tasks. It gets its name from the term *do-file*, which is a file containing a list of commands for Stata to run (what is called a batch file in other settings). Although it has advanced features that can help in writing such files, it can also be used to build up series of commands that can then be submitted to Stata all at once. This feature can be handy when writing a loop to process multiple variables in a similar fashion or when doing complex repetitive tasks interactively.

To get the most from this chapter, you should work through it at your computer.

The Do-file Editor toolbar

The Do-file Editor has 13 buttons. Many of the buttons have a similar purpose as their lookalikes in the main Stata toolbar.

If you ever forget what a button does, hold the mouse pointer over a button for a moment, and a tooltip will appear with a description of that button.

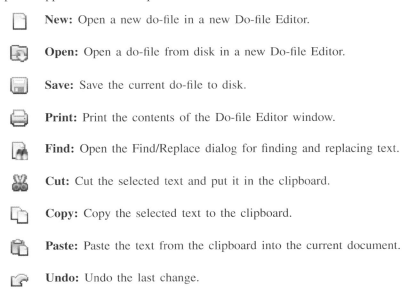

New: Open a new do-file in a new Do-file Editor.

Open: Open a do-file from disk in a new Do-file Editor.

Save: Save the current do-file to disk.

Print: Print the contents of the Do-file Editor window.

Find: Open the Find/Replace dialog for finding and replacing text.

Cut: Cut the selected text and put it in the clipboard.

Copy: Copy the selected text to the clipboard.

Paste: Paste the text from the clipboard into the current document.

Undo: Undo the last change.

Redo: Undo the last undo.

Preview: Open a Viewer window to display the contents of the Do-file Editor window. This feature is particularly useful when editing a SMCL file.

Run: Run the commands in the do-file without showing any output. If text is highlighted, the button becomes the **Run Selected Lines** button and will run just the selected lines, suppressing all output. This feature is useful when assembling long series of commands.

 Do: Run the commands in the do-file, showing all commands and their output. If text is highlighted, the button becomes the **Do Selected Lines** button and will run only the selected lines, showing all output.

Using the Do-file Editor

Suppose that we would like to analyze fuel usage for 1978 automobiles similar to that done in the sample session of chapter 3. We know that you will be issuing many commands to Stata during your analysis and that we want to be able to reproduce your work later without having to type each command again.

We can do this easily in Stata: simply save a text file containing the commands. When this is done, we can tell Stata to run the file and execute each command in sequence. Such a file is known as a Stata "do-file"; see [U] **16 Do-files**.

To analyze fuel usage of 1978 automobiles, we would like to create a new variable giving gallons per mile. We would like to see how that variable changes in relation to vehicle weight for both domestic and imported cars. Performing a regression with our new variable would be a good first step.

To get started, click the **Do-file Editor** button to open the Do-file Editor. After the Do-file Editor opens, type the commands below, since they are what we would like to submit. Purposely misspell the name of the `foreign` variable on the fourth line. (We are intentionally making some common mistakes and then pointing you to the solutions. This will save you time later.)

```
sysuse auto
generate gpm = 1/mpg
label var gpm "Gallons per mile"
regress gpm weight foreing
```

Here is what your Do-file Editor should look like now:

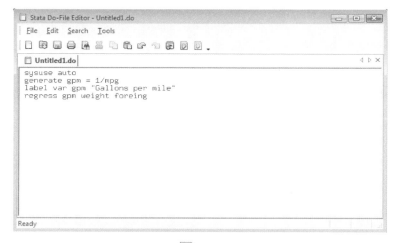

Click the **Do** button to run the commands: .

When you click the **Do** button, Stata executes the commands in sequence, and the results appear in the Results window:

```
. do "C:\Users\mydir\AppData\Local\Temp\STD08000000.tmp"

. sysuse auto
(1978 Automobile Data)

. generate gpm = 1/mpg

. label var gpm "Gallons per mile"

. regress gpm weight foreing
variable foreing not found
r(111);

end of do-file

r(111);
```

The do "C:\..." is how Stata executes the commands in the Do-file Editor. Stata saves the commands to a temporary file and issues the do command to execute them.

Everything worked as planned until Stata saw the misspelled variable. The first three commands were executed, but an error was produced on the fourth. Stata does not know of a variable named foreing. We need to go back to the Do-file Editor and correct the mistake.

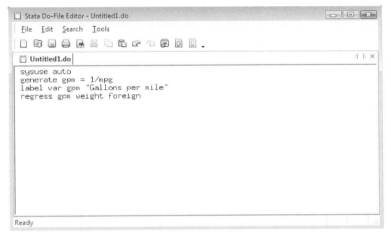

We click the Run button again. Alas, Stata now fails on the first line—it will not overwrite the dataset in memory that we had changed.

```
. do "C:\Users\mydir\AppData\Local\Temp\STD08000000.tmp"

. sysuse auto
no; data in memory would be lost
r(4);

end of do-file

r(4);
```

We now have a choice for what we should do:

- We could use our do-file to automatically clear out Stata's memory before it runs. Doing so is convenient, but dangerous, because it defeats Stata's protection against throwing away changes without warning.
- We could manually clear the dataset and then process the do-file again. This process can be aggravating when building a complicated do-file.

Here is some advice: automatically clear Stata's memory while debugging the do-file. Once the do-file is in its final form, decide the context in which it will be used. If it will be used in a highly automated environment (such as when certifying), the do-file should still automatically clear Stata's memory. If it will be used rarely, do not clear Stata's memory. It will save heartache.

We will add a `clear` option to the `sysuse` command to automatically clear the dataset in Stata's memory before the do-file runs:

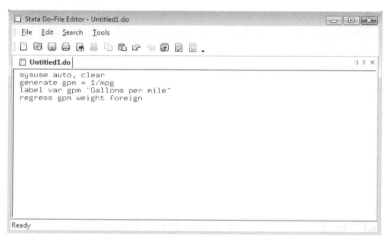

The do-file now runs well, as clicking on the Do button shows:

```
. do "C:\Users\mydir\AppData\Local\Temp\STD08000000.tmp"

. sysuse auto, clear
(1978 Automobile Data)

. generate gpm = 1/mpg

. label var gpm "Gallons per mile"

. regress gpm weight foreign

      Source |       SS       df       MS              Number of obs =      74
-------------+------------------------------           F(  2,    71) =  113.97
       Model |  .009117618      2   .004558809          Prob > F      =  0.0000
    Residual |   .00284001     71      .00004           R-squared     =  0.7625
-------------+------------------------------           Adj R-squared =  0.7558
       Total |  .011957628     73   .000163803          Root MSE      =  .00632

------------------------------------------------------------------------------
         gpm |      Coef.   Std. Err.      t    P>|t|     [95% Conf. Interval]
-------------+----------------------------------------------------------------
      weight |   .0000163   1.18e-06    13.74   0.000     .0000139    .0000186
     foreign |   .0062205   .0019974     3.11   0.003     .0022379    .0102032
       _cons |  -.0007348   .0040199    -0.18   0.855    -.0087504    .0072807
------------------------------------------------------------------------------

.
end of do-file
```

You might want to select **File > Save As...** from the Do-file Editor to save this do-file. Later, you could select **File > Open...** to open it and then add more commands as you move forward with your analysis. By saving the commands of your analysis in a do-file as you go, you do not have to worry about retyping them with each new Stata session. Think hard about removing the clear from the first command.

After you have saved your do-file, you can execute the commands it contains by typing do *filename*, where the *filename* is the name of your do-file.

The File menu

The **File** menu of the Do-file Editor includes standard features found in most text editors. You may choose to start a **New** file, **Open...** an existing file, **Save** the current file, or save the current file under a new name with **Save As...**. You may also **Print...** the current file. There are also buttons on the Do-file Editor's toolbar that correspond to these features.

There is one other useful feature under the **File** menu: you may select **Insert File...** to insert the contents of another file at the current cursor position in the Do-file Editor.

Editing tools

The **Edit** menu of the Do-file Editor includes the standard **Cut**, **Copy**, and **Paste** capabilities, along with a multilevel **Undo** and **Redo**. There are also buttons on the Do-file Editor's toolbar for easy access to these capabilities. There are several other **Edit** features that you may find useful.

- You may delete or select the current line.
- You may also shift the current line or selection right or left one tab stop.

- • The **Change Case** menu item allows you to change the case of the character to the right of the cursor or to change an entire selection. When you **Change Case** of an entire selection, the editor will compare the number of lowercase characters with the number of uppercase characters. The case with the greater number of characters will be changed. For example, if you have aBcDe selected, you will have ABCDE after selecting **Change Case**. The editor sees that there are fewer uppercase (2) than lowercase (3) characters, so it makes the entire selection uppercase. If you select **Change Case** again, the entire selection will change to lowercase. There are fewer lowercase (0) than uppercase (5) characters, so all are switched to lowercase. If there is no selection when you select **Change Case**, the case of the character immediately to the right of the cursor is switched. The cursor is also moved one character to the right. For example, if the cursor is to the left of 'a' in aBcDe when you select **Change Case**, you will see ABcDe and the cursor will be just to the left of 'B'. If you select **Change Case** again, you will see AbcDe, and the cursor will be just to the left of 'c'.

Preferences

When you right-click on the Do-file Editor window and select **Preferences...** from the contextual menu, you may customize the way the editor behaves.

You may set the number of spaces each tab character represents. You may also set whether the Do-file Editor automatically indents the current line to the same level of indentation as the previous line.

You may also change the font and font size, and you may set whether or not the Do-file Editor automatically saves the current file when you click the **Do** button or the **Run** button. Normally, the file on which you are working is not saved to disk unless you explicitly select **File > Save** or **File > Save As...**. For example, if you open C:\data\myfile.do and make some changes, those changes will not be saved until you save them from the **File** menu. The changes are not saved, even when you execute myfile.do by clicking the **Do** button or the **Run** button. This prevents your original file from being overwritten until you tell the Do-file Editor to overwrite it. If you check *Auto-save on Do/Run* in the*Do-file Editor Preferences* dialog, any changes that you have made to myfile.do will be saved when you click either the **Do** button or the **Run** button.

Searching

Stata's Do-file Editor includes standard **Find...** and **Replace...** capabilities in the **Search** menu. You should already be familiar with these capabilities from other text editors and word processors. You may access the **Find** capability from a button on the toolbar as well.

The **Search** menu also includes some choices that are useful when you are writing a long do-file or ado-file. You may select **Search > Go to Line...** and jump straight to any line in your file.

Matching and balancing of parentheses ( ), braces { }, and brackets [ ] are available from the **Search** menu. When you select **Search > Match**, the Do-file Editor looks at the character immediately to the right of the cursor. If it is one of the characters that the editor can match, the editor will find the matching character and place the cursor immediately in front of it. If there is no match, you will hear a beep and the cursor will not move.

When you select **Search > Balance**, the Do-file Editor looks to the left and right of the current cursor position or selection and creates a selection including the narrowest level of matching characters. If you select **Balance** again, the editor will expand the selection to include the next level of matching characters. If there is no match, you will hear a beep and the cursor will not move.

Balance is easier to explain with an example. Type (now (is the) time) in the Do-file Editor. Place the cursor between the words is and the. Select **Search > Balance**. The Do-file Editor will select (is the). If you select **Balance** again, the Do-file Editor will select (now (is the) time).

The Tools menu

You have already learned about the **Do** button. Clicking it is equivalent to choosing **Tools > Do** in the Do-file Editor.

Next to the **Do** button is the **Run** button. Clicking it is equivalent to choosing **Tools > Run**. **Run** executes all the commands in the Do-file Editor just like **Do** does, but **Run** suppresses output. It is unlikely that you will ever need to use **Run**; see [U] **16.6.2 Suppressing output** for more details.

Do and **Run** are equivalent to Stata's do and run commands; see [U] **16 Do-files** for a complete discussion.

There are two other useful choices on the **Tools** menu. If you wish to execute a subset of the lines in your do-file, highlight those lines, and select **Tools > Do Selection**.

If you wish to execute all commands from the current line through the end of the file, select **Tools > Do to Bottom**.

You can also preview files in the Viewer by selecting **Tools > Preview File in Viewer** or by clicking the **Preview File in Viewer** button. This feature is useful when previewing a SMCL file—which will happen when you write your own help files.

Saving interactive commands from Stata as a do-file

While working interactively with Stata, you may decide that you would like to rerun the last several commands that you typed interactively. You can save the contents of the Review window as a do-file and open that file in the Do-file Editor. You can copy a command from a dialog (rather than submit it) and paste it into the Do-file Editor. See chapter 4 for details. Also see [R] **log** for information on the cmdlog command, which allows you to log all commands that you type in Stata to a do-file.

Notes

15 Graphing data

Working with graphs

Stata has a rich system for graphical representation of data. The main command for creating graphs is unsurprisingly named `graph`. Behind this plain name is a wealth of tools. In this chapter, we will make one simple graph to point out the basics of the Graph window.

A simple graph example

In the sample session of chapter 3, we made a scatterplot, added a fitted regression line, and made a grid of scatterplots to allow comparisons across groups. Here we make a simple box plot that shows the displacements of the cars' engines and how they compare across repair records within the place of manufacture of the cars, using the `auto` dataset (`sysuse auto`).

We select **Graphics > Box plot**, choose `displacement` in the *Variables* field on the **Main** tab, click the **Categories** tab, check the *Group 1* checkbox and enter `rep78` for the first group, and check the *Group 2* checkbox and enter `foreign` for the second group. Finally, we click the **Submit** button, so that we could make changes to the graph if need be. After we look at the graph, we realize that we forgot the title. We close the Graph window, click on the **Titles** tab of the *graph box* dialog, add the title *Displacement across Repairs within Origin*, and click the **Submit** button again.

The Graph window comes up, showing us the fruit of our labor:

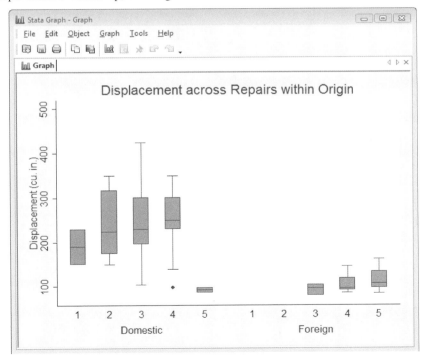

Graph window

When the Graph window comes up, it shows our graph in a window with a toolbar. The first four icons are familiar to us from other Stata windows: **Open**, **Save**, **Print**, and **Copy**. The next two icons are new:

 Rename Graph: This allows the graph to be renamed. Why would you do this? If you would like to have multiple graphs open at once, the graphs need to be named. So, you can click the **Rename** button to give a graph a name. This window will then remain open when you create your next graph.

 Graph Editor: Stata has a Graph Editor that allows you to manipulate and edit your graph. This feature will be introduced in the next chapter.

The inactive buttons to the right of the **Start Graph Editor** button are used by the Graph Editor, so their meanings will become clear in the next chapter.

We decide that we like this graph and would like to save it. We can save it either by clicking on the **Save** button and choosing a name and a location or by right-clicking on the Graph window itself, and selecting **Save Graph...**.

Saving and printing graphs

You can save a graph once it is displayed by right-clicking on its window and selecting **Save Graph...**. You can print a graph by right-clicking on its window and selecting **Print...**. You can also use the **File** menu to save or print a graph. We recommend that you always right-click on a graph to save or print it to ensure that the correct one is selected.

For more information about printing graphs, see the *Stata Graphics Reference Manual*.

Right-clicking on the Graph window

Right-clicking on the Graph window displays a menu from which you can select
- **Save Graph...** to save the graph to disk
- **Copy Graph** to copy the graph to the Clipboard
- **Start Graph Editor...** to start the Graph Editor
- **Preferences...** to edit the preferences for graphs
- **Print...** to print the contents of the Graph window

The Graph button

The **Graph** button is located on the main window's toolbar. The button has two parts, an icon and an arrow. Clicking on the icon brings the topmost Graph window to the front of all other windows. Clicking on the arrow displays a menu of open graphs. Selecting a graph from the menu brings the graph to the front of all other windows. If you ever decide to close the Graph window, you can reopen it only by reissuing a Stata command that draws a new graph.

16 Editing graphs

Working with the Graph Editor

With Stata's Graph Editor you can change almost anything on your graph, and you can add text, lines, arrows, and markers wherever you would like.

We will first make a graph to edit and will then point out the tools in the Graph Editor. Here is the command that we'll use to make the graph:

```
. scatter mpg weight, name(mygraph) title(Mileage vs. Vehicle Weight)
```

Start the editor by right-clicking on your graph and selecting **Start Graph Editor**. Click once on the title of the graph. Here is a picture of the Graph Editor with its elements labeled.

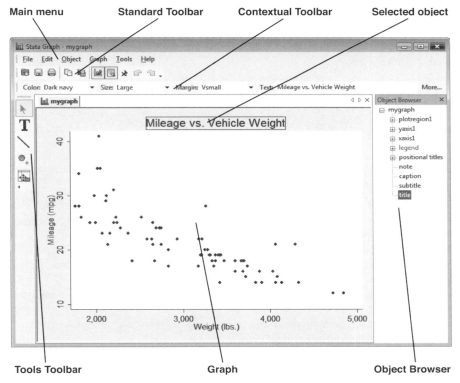

Select any of the tools along the left of the editor to edit the graph. The Pointer (Select Tool) is selected by default.

You can change the properties of objects or drag them to new locations by using the Pointer. As you select objects with the Pointer, a Contextual Toolbar will appear just above the graph. In our example, the title of the graph is selected, so the Contextual Toolbar has controls that are relevant for editing titles. You can use any of the controls on the Contextual Toolbar to immediately change the most important properties of the selected object. Right-click on an object to access more properties and operations. Hold the **Shift** key when dragging objects to constrain the movement to horizontal, vertical, or 90-degree angles.

Add text, lines, or markers (with optional labels) to your graph by using the three *Add...* tools— **T** , $\searrow$, and $\bullet_{+}$. Lines can be changed to arrows by using the Contextual Toolbar. If you do not like the default properties, simply change their settings in the Contextual Toolbar before adding the text, line, or marker. The new setting will then be applied to all added objects, even in future Stata sessions.

Do not be afraid to try things. If you do not like a result, change it back by using the same tool, or click the Undo button, ☞ , in the Standard Toolbar for the Graph Editor (below the main menu). **Edit > Undo** in the main menu does the same thing.

Remember to reselect the Pointer tool when you want to drag objects or change their properties.

You can move objects on the graph and have the rest of the objects adjust their position to accommodate the move with the Grid Edit tool, ⊕ . With this tool, you are repositioning objects in the underlying grid that holds the objects in the graph. Some graphs, e.g., `by graphs`, are composed of nested grids. You can reposition objects only within the grid that contains them; they cannot be moved to other grids.

You can also select objects in the Object Browser along the right of the graph. This window shows a hierarchical listing of the objects in the graph. Clicking or right-clicking on an object in the browser is the same as clicking or right-clicking on the object in the graph.

Stop the editor by selecting **File > Stop Graph Editor** from the main menu or by clicking the Graph Editor button. When you stop the Graph Editor, you will be prompted to save your graph if you have made any changes. If you do not save your graph, your changes will not be lost, but you will risk losing them if you create a new graph in the same Graph window. You must stop the editor if you would like to work on other tasks in Stata.

Here are a few of the things that you can do with the editor.

- Add annotations using lines, arrows, and text.
- Add or remove grid lines or reference lines.
- Add or modify titles, captions, and notes.
- Change scatterplots to line plots, connected plots, areas, bars, spikes, or drop lines—and, of course, vice versa.
- Change the size, color, margin, and other properties of your graph's titles (or any other text on the graph).
- Move your legend to another side of the graph, or even place it in the plot region.
- Change the aspect ratio of your graph.
- Stack the bars on a bar graph or turn them into percentages.
- Rotate or change the angle of axis labels.
- Add custom ticks and labels to the axes.
- Change the rule for the number and spacing of ticks and labels on an axis.
- Emphasize a point on the graph, whether marker, bar, spike, or other plot, by making it a custom color, size, or symbol.
- Change the text or properties of a marker label.

Since you can edit every property of every object on the graph, you can change almost anything about your graph. To learn more, see [G] **graph editor** or type `help graph editor`.

17 Saving and printing results by using logs

Using logs in Stata

When working on an analysis, it is worthwhile to behave like a bench scientist and keep a lab notebook of your actions so that your work can be easily replicated. Everyone has a feeling of complete omniscience while working intensely—this feeling is wonderful but fleeting. The next day, the exact small details needed for perfect duplication have become obscure. Stata has exactly such a lab notebook at hand: the *log* file.

A log file is simply a record of your Results window. It records all commands and all textual output as it happens. Thus, it keeps your lab notebook for you as you work. Since it writes the file as it writes the Results window, it also protects you from disastrous failures, be they power failures or computer crashes. We recommend that you start a log file whenever you begin any serious work in Stata.

Logging output

All the output that appears in the Results window can be captured in a log file. Stata can save the file in one of two different formats. By default, Stata will save the file in its Stata Markup and Control Language (SMCL) format, which preserves all the formatting and links from the Results window. You can open these results in the Viewer and they will behave as though they were in the Results window. If you would rather have plain-text files without any formatting, you can save the file as a plain *log* file. We recommend using the SMCL format, because SMCL files can be translated into a variety of formats readable by applications other than Stata with the **File > Log > Translate...** menu (see [R] **translate**).

To start a log file, click the **Log** button, . Doing so will open a standard file dialog allowing you to specify a directory and filename to save your log. If you do not specify a file extension, the extension .smcl will be added to the filename. If you specify a file that already exists, you will be asked whether you want to append the new log to the file or overwrite the file with the new log.

Here is an example of a short session. There are a few items of interest.

- The header showing the log file's location, type, and starting timestamp is part of the log file. This feature helps when working with multiple log files.
- The two lines starting with asterisks (*) are *comments*. Stata ignores the text following the asterisk, so you may type any comment you would like, with any special characters you would like. Commenting is a good way to document your thought process and to mark sections of the log for later use.
- In the example below, the log file was closed using the `log close` command. Doing so is not strictly necessary because log files are automatically closed when you exit Stata.

113

```
. log using C:\data\base.smcl

      log:  C:\data\base.smcl
 log type:  smcl
opened on:  15 Apr 2007, 15:47:01
. sysuse auto
(1978 Automobile Data)
. by foreign, sort: summarize price mpg

-> foreign = Domestic
    Variable |      Obs        Mean    Std. Dev.       Min         Max

       price |       52    6072.423     3097.104      3291       15906
         mpg |       52    19.82692     4.743297        12          34

-> foreign = Foreign
    Variable |      Obs        Mean    Std. Dev.       Min         Max

       price |       22    6384.682     2621.915      3748       12990
         mpg |       22    24.77273     6.611187        14          41
. * be sure to include the above stats in report!
. * now for something completely different
. corr price mpg
(obs=74)

             |    price      mpg

       price |   1.0000
         mpg |  -0.4686    1.0000
. log close
      log:  C:\data\base.smcl
 log type:  smcl
closed on:  15 Apr 2007, 15:47:01
```

What was not shown here is that Stata allows multiple log files to be open at once. For details on this, see `help log`.

Working with logs

Log files are best viewed using Stata's Viewer. Select **File > Log > View...** If there is a log file open (as shown by the status bar), it will be the default log file to view; otherwise, you need to either type the name of the log file into the dialog or click the **Browse...** button to find the file with a standard file dialog.

Once you are in the Viewer window, everything behaves as expected: you can copy text and paste between the Viewer and anything else that uses text, such as word processors or text editors. You can even paste into the Command window or the Do-file Editor, but you should take care to copy only commands, not their output or the prompt ("`.`") at the start of the echoed command. When working with a word processor, what you paste will be unformatted text; it will look best if you use a fixed-width font like Courier to display it.

Viewing your current log file is a good way to keep a reminder of something you have already done or a view of a previous result. The Viewer window takes a snapshot of your log file and hence will not scroll as you keep working in Stata. If you need to see more recent results in the Viewer, press the **Refresh** button.

For more detailed information about logs, see [U] **15 Printing and preserving output** and [R] **log**. For more information about the Viewer, see chapter 5.

Printing logs

To print a standard SMCL log file, you need to have the log file open in a Viewer window. Once the log file is in the Viewer, you can either right-click on the Viewer window, and select **Print...**, or select the Viewer window from **File > Print >** *"viewer title"*. A *Print* dialog will appear. After you click **Print**, an *Output Settings* dialog will appear.

- You can fill in none, any, or all of the items *Header*, *Name*, and *Project*. You can check or uncheck options to *Print Line #s*, *Print Header*, and *Print Logo*. These items are saved and will appear again in the *Print Settings* (in this and in future Stata sessions).

- You can set the margins and color scheme that the printer will use by clicking on **Prefs...** in the *Output Settings* dialog to open the *Printer Preferences* dialog. Monochrome is for black-and-white printing, Color is for default color printing, and Custom 1 and Custom 2 are for customized color printing.

- You can set the font that the printer will use from the *Printer Preferences* dialog. The font dialog will list only the fixed-width "typewriter" fonts (e.g., Courier) available for your printer. Stata, by default, will choose a font size that it thinks is appropriate for your printer.

You could also use the `translate` command to generate a Postscript file. See [R] **translate** for more information.

If your log file is a plain-text file (`.log` instead of `.smcl`), you can load it into a text editor, such as Notepad, the Do-file Editor in Stata, or your favorite word processor. You can then edit the log file—add headings, comments, etc.—format it, and print it. If you bring the log file into a word processor, it will be displayed and printed with its default font. The log file will not be easily readable when printed in a proportionally spaced font (e.g., Times Roman or Helvetica). It will look much better printed in a fixed-width font (e.g., Courier).

You may wish to associate the `.log` extension with a text editor (such as Notepad, Write, or WordPad) in Windows. You can then edit and print the logs from those Windows applications if you like.

(Continued on next page)

Rerunning commands as do-files

Stata can also log just the commands from a session without recording the output. This feature is a convenient way to make a do-file interactively. Such a file is called a *cmdlog* file by Stata. You can start a `cmdlog` file by typing

cmdlog using *filename*,

and you can close the `cmdlog` file by typing

cmdlog close

Here, for example, is what a `cmdlog` of the previous session would look like. It contains only commands and hence could be used as a do-file.

```
sysuse auto
by foreign, sort: summarize price mpg
* be sure to include the above stats in report!
* now for something completely different
corr price mpg
```

If you start working and then wish you had started a `cmdlog` file, you can save yourself heartache by saving the contents of the Review window. The Review window stores the last 5,000 commands you typed. Simply right-click on the Review window and select **Save Review Contents...** from the menu. If you would like to move the commands directly to the Do-file Editor, select **Copy Review Contents to Do-file Editor**. You may find this method a more convenient way to create a text file containing only the commands that you typed during your session.

See chapter 14, [U] **16 Do-files**, and [U] **15 Printing and preserving output** for more information.

18 Setting font and window preferences

Changing and saving fonts and sizes and positions of your windows

You may find that you would like to change the fonts and display style of Stata's windows, depending on your monitor resolution and personal preferences. At the same time, there could be requirements for font usage, say, when you submit graphs to journals. Stata accommodates both of these by allowing sets of preferences for how windows are displayed.

We'll first cover what can be changed in each window and then talk about what you can manage with your preferences.

Graph window

The preferences for the Graph window can be changed by right-clicking on the Graph window and choosing **Preferences...** from the contextual menu. The fonts that can then be set for graphs as they display in Stata (under the **Window**) tab. The fonts that should be used when printing can be set under the **Printer** tab. The behavior of the clipboard is controlled under the **Clipboard** tab.

The Graph preferences also allow various *schemes*. These provide a quick way to optimize graphs for printing or display on a screen. There are even schemes defined for the *The Economist* and the *Stata Journal* so that you can get the details for these journals right without much fuss. Changing the scheme *does not* change the current graph—it applies the settings to future graphs.

All other windows

You can change the display font and font size for all other types of windows in Stata.

Fonts and font sizes can be changed for any type of window by right-clicking the window and selecting **Font...** Doing so will bring up the Font dialog, from which you can pick the font and size of your choice. The font lists for each of the Results, Viewer, Data Editor, and Do-file Editor windows are restricted to fixed-width fonts only. This restriction ensures that output and numbers line up properly and are readable. The Command, Variables, and Review windows can have any font that you would like.

Changing color schemes

The Results and Viewer windows have color schemes that control the way in which input, text, results, errors, links, and highlighted text display. Each has its color scheme set in the same fashion: you can right-click on the window and select or design your own color scheme. The default setting for the Results window is the built-in Black Background scheme, whereas the default setting for the Viewer is the built-in White Background scheme. There is also a Blue Background scheme for those nostalgic for DOS. Finally, there are three settings for custom schemes, each of which starts as one of the three built-in schemes. If you want your Viewer to look exactly like a Results window, set them to have the same scheme. The settings for the Viewer affect all Viewer windows at once.

Managing multiple sets of preferences

Stata's preferences are automatically saved when you exit Stata, and they are reloaded when Stata is launched. However, sometimes you may wish to rearrange Stata's windows and then revert the windows to your preferred arrangement. You can do this by saving your preferences to a named preference set and loading them later. Any changes you make to Stata's preferences after loading a preferences set do not affect the set; the set remains untouched unless you specifically overwrite it.

To save and load preferences, open the **Edit > Preferences > Manage Preferences** menu, and do the following:

- Select a preference set from the **Load Preferences** menu to load it.
- Select **Load Preferences > Factory Settings** to restore the preferences to their factory settings. Other default windowing settings can be found in the **Load Preferences** menu.
- Select **Save Preferences > New Preferences Set...** to save the current preferences to a set. Enter a name for the set, and click **OK**.
- Select an existing set from the **Save Preferences** menu to overwrite it with the current preferences.
- Select a preference set from the **Delete Preferences** menu to delete it. Click **OK** to verify that you wish to delete the set.

Closing and opening windows

You can close all windows but the Results and Command windows. If you want to open a closed window, open the **Window** menu and select the desired window.

19 Learning more about Stata

Where to go from here

You now know plenty enough to use Stata. There is still much, much more to learn, because Stata is a rich environment for doing statistical analysis and data management. What should you do to learn more?

1. Get an interesting dataset and play with Stata.
   a. Use the menus and dialog system to experiment with commands. Notice what commands show up in the Results window. You will find that Stata's simple and consistent command syntax will make the commands easy to read, so that you'll know what you've done, and easy to remember, so that typing some commands will be faster than using menus.
   b. Play with graphs and the Graph Editor.
2. If you venture into the Command window, you will find that many things will go faster. You will also find that it is possible to make mistakes where you cannot understand why Stata is balking.
   a. Try `help` *commandname* or **Help > Stata Command...** and entering the command name.
   b. Look at the command syntax, and the examples in the help file, and compare them with what you typed. Compare them closely: small typos make commands impossible for Stata to parse.
3. Explore Stata by selecting **Help > Search...** and choosing **Search documentation and FAQs**. You will uncover many statistical routines that could be of great use.
4. Look through the Combined Subject Table of Contents in the *Stata Quick Reference and Index*.
5. Read and work your way through the *User's Guide*. It is designed to be read cover to cover, and it contains most of the information you need to become an expert Stata User. It is well worth reading.
6. Flip through the reference manuals to read about statistical methods you like to use. The reference manuals are not meant to be read cover to cover—they are meant to be read like an encyclopedia. You can find the datasets used in the examples in the manuals at http://www.stata-press.com/data/. Doing so will enable you to work through the examples quickly.
7. Stata has much information, including answers to frequently asked questions (FAQs), at http://www.stata.com/support/faqs/.
8. There are many useful links to Stata resources at http://www.stata.com/links/resources.html.
9. If you prefer to sample the *User's Guide* and the references, there is some advice later in this chapter for you.
10. Take a Stata NetCourse™. NetCourse 101 is an excellent choice to learn about Stata. See http://www.stata.com/netcourse/ for course information and schedules.

Suggested reading from the User's Guide and reference manuals

The *User's Guide* is designed to be read from cover to cover. The reference manuals are designed as references to be sampled when necessary.

Ideally, after reading this *Getting Started* manual, you should read the *User's Guide* from cover to cover, but you probably want to become at least somewhat proficient in Stata right away. Here is a suggested reading list of sections from the *User's Guide* and the reference manuals to help you on your way to becoming a Stata expert.

This list covers fundamental features and points you to some less obvious features that you might otherwise overlook.

Basic elements of Stata

[U] Chapter 11   Language syntax

[U] Chapter 12   Data

[U] Chapter 13   Functions and expressions

Memory

[U] Chapter 6   Setting the size of memory

[D] compress   Compress data in memory

Data input

[U] Chapter 6   Setting the size of memory

[U] Chapter 21   Inputting data

[D] edit   Edit and list data with Data Editor

[D] infile   Quick reference for reading data into Stata

[D] insheet   Read text (ASCII) data created by a spreadsheet

[D] append   Append datasets

[D] merge   Merge datasets

Graphics

Stata Graphics Reference Manual

Useful features that you might overlook

[U] Chapter 28   Using the Internet to keep up to date

[U] Chapter 16   Do-files

[U] Chapter 19   Immediate commands

[U] Chapter 23   Dealing with strings

[U] Chapter 24   Dealing with dates

[U] Chapter 25   Dealing with categorical variables

[U] Chapter 13   Accessing coefficients and standard errors

[U] Chapter 13   Accessing results from Stata commands

[U] Chapter 26   Overview of Stata estimation commands

[U] Chapter 20   Estimation and postestimation commands

[U] Chapter 13   Accessing coefficients and standard errors

[R] estimates   Save and manipulate estimation results

Basic statistics

[R] anova   Analysis of variance and covariance

[R] ci   Confidence intervals for means, proportions, and counts

[R] correlate   Correlations (covariances) of variables or estimators

[D] egen   Extensions to generate

[R] regress   Linear regression

[R] predict   Obtain predictions, residuals, etc., after estimation

[R] regress postestimation   Postestimation tools for regress

[R] test	Test linear hypotheses after estimation
[R] summarize	Summary statistics
[R] table	Tables of summary statistics
[R] tabulate oneway	One-way tables of frequencies
[R] tabulate twoway	Two-way tables of frequencies
[R] ttest	Mean-comparison tests

Matrices

| [U] Chapter 14 | Matrix expressions |
| [U] Chapter 18 | Scalars and matrices |

Mata Reference Manual

Programming

[U] Chapter 16	Do-files
[U] Chapter 17	Ado-files
[U] Chapter 18	Programming Stata
[R] ml	Maximum likelihood estimation

Stata Programming Reference Manual

Mata Reference Manual

System values

| [R] set | Overview of system parameters |
| [P] creturn | Return c-class values |

Internet resources

The Stata web site (http://www.stata.com/) is a good place to get more information about Stata. Half of the web site is dedicated to user support. You will find answers to FAQs, ways to interact with other users, official Stata updates, and other useful information. You can also subscribe to Statalist, a listserver devoted to Stata and statistics discussion.

You will also find information on Stata NetCourses™, which are interactive courses offered over the Internet and vary in length from a few weeks to 8 weeks. Visit the Stata web site for more information.

At the web site is the Stata Bookstore, which contains books that we feel may be of interest to Stata users. Each book has a brief description written by a member of our technical staff explaining why we think this book may be of interest.

If you have access to the web, we suggest that you take a quick look at the Stata web site now. You can register your copy of Stata online and request a free subscription to the *Stata News*.

Visit http://www.stata-press.com/ for information on books, manuals, and journals published by Stata Press. The datasets used in examples in the Stata manuals are available from the Stata Press web site.

Also visit http://www.stata-journal.com/ to read about the *Stata Journal*, a quarterly publication containing articles about statistics, data analysis, teaching methods, and effective use of Stata's language.

See chapter 20 for details on accessing official Stata updates and free additions to Stata on the Stata web site.

Notes

20 Updating and extending Stata—Internet functionality

Internet functionality in Stata

Stata works well together with the Internet. It can use datasets and view remote help files as though they were on your computer. It can keep itself up to date (with your permission, of course). Finally, you can install *user-written commands*, which are commands that extend Stata's functionality. These are commands that have been presented in the *Stata Journal* (SJ), the *Stata Technical Bulletin* (STB), or have simply been written and shared by the greater Stata community.

This chapter will show you how you can expand Stata's horizons.

Using files from the Internet

Stata understands URLs as though they were local file locations. If you know of a file on the web that you would like to use, be it a dataset, a graph, or a do-file, you can open it in Stata easily. Here is a small example.

There are many datasets at http://www.stata-press.com/data/. Suppose that you would like to use the census12 dataset used in [U] **11 Language syntax**, and you know that its location is at http://www.stata-press.com/data/r10/census12.dta. Since you know that the command for opening a dataset is use, you could do the following:

```
. use http://www.stata-press.com/data/r10/census12.dta
(1980 Census data by state)

. describe
Contains data from http://www.stata-press.com/data/r10/census12.dta
  obs:            50                          1980 Census data by state
 vars:             6                          18 Mar 2007 06:01
 size:         2,050  (99.9% of memory free)

                storage  display     value
variable name     type   format      label      variable label

state            str14   %14s                    State
region           str7    %9s                     Census region
pop              long    %10.0g                  Population
median_age       float   %9.2f                   Median age
marriage_rate    float   %9.0g
divorce_rate     float   %9.0g

Sorted by:
```

This functionality is everywhere in Stata: any command that reads a file via a *filename* in its syntax can use a web address as easily as a file that is stored on your computer.

Official Stata updates

By official Stata, we mean the pieces of Stata that are provided and supported by StataCorp. The other and equally important pieces are the user-written additions published in the SJ, distributed over Statalist, or distributed in other ways.

Stata can fetch both official updates and user-written programs from the Internet. In the first case, you choose **Help > Official Updates**. In the second case, you choose **Help > SJ and User-written Programs**. (There are commands for doing this, too; see [U] **28 Using the Internet to keep up to date**.)

Let's start with the official updates. There are two parts to official Stata: the Stata executable and Stata's ado-files.

Important note for Windows Vista users: you must start Stata in a special fashion to install official updates. Right-click on the Stata icon, and select **Run as Administrator** from the contextual menu. You must do this even if your account is an administrator's account. If you do not run Stata as an administrator, Vista will not allow you to install any of the files needed to update Stata.

StataCorp often releases updates to official Stata. These updates are to add new features and, sometimes, to fix bugs. Typically the updates are to the ado-files because, in fact, most of Stata is written in Stata's ado-language. Occasionally, we update the executable as well.

When you choose **Official Updates**, Stata tells you the dates of its two official pieces:

```
update

Stata executable
    folder:                 C:\Program Files\Stata10\
    name of file:           wsestata.exe
    currently installed:    28 Jul 2007
Ado-file updates
    folder:                 C:\Program Files\Stata10\ado\updates\
    names of files:         (various)
    currently installed:    28 Jul 2007
Recommendation
    compare these dates with what is available from
        http://www.stata.com
        other location of your choosing
        cdrom drive
```

If you know of a location that has official updates other than the Stata web site, click on `other location of your choosing`, and fill in the site or directory name that you wish.

If you are connected to the Internet, you can click on `http://www.stata.com` to compare the dates of your files with the latest updates available on the Stata web site. If you have trouble connecting to the Internet from Stata, visit http://www.stata.com/support/faqs/web/ for help.

When you click on `http://www.stata.com`, Stata compares your two official pieces with the most up-to-date versions available:

```
update query

(contacting http://www.stata.com/)
Stata executable
    folder:                C:\Program Files\Stata10\
    name of file:          wsestata.exe
    currently installed:   28 Jul 2007
    latest available:      28 Jul 2007
Ado-file updates
    folder:                C:\Program Files\Stata10\ado\updates\
    names of file:         (various)
    currently installed:   28 Jul 2007
    latest available:      28 Jul 2007
Recommendation
    Do nothing; all files up to date.
```

Here all files are up to date.

You might be told that you need to update your ado-files, your executable, or both. Entering `update all` will handle all cases. The sections **Updating the official ado-files** and **Updating the executable** are for informational purposes only. You should never update one without also updating the other if both updates are available. An update is not complete until you follow all the instructions presented on the screen.

Automatic update checking

Stata can periodically check for updates for you. By default, Stata will check once every 7 days for updates from Stata's web site. The 7-day interval is from the last time an `update query` was performed regardless of whether it was by Stata or by you. You can change the interval between checks.

Before Stata connects to the Internet to check for an update, it will ask you if you would like to check now, check the next time Stata is launched, or check after the next interval. You can disable the prompt and allow Stata to check without asking.

If an update is available, Stata will notify you. From there, you should follow the recommendations for updating Stata.

You can change the settings for automatic update checking by selecting **Edit > Preferences > General Preferences...** and choosing **Internet**.

Updating the official ado-files

Much of Stata is implemented as ado-files—text files containing programs in Stata's ado-language. StataCorp periodically releases updates. The **Official Updates** system makes it easy to keep your official ado-files up to date.

After selecting **Help > Official Updates** and then clicking on `http://www.stata.com`, you might see

---

update query

(contacting http://www.stata.com/)

Information about other types of updates may appear . . .

```
Ado-file updates
    folder:               C:\Program Files\Stata10\ado\updates\
    names of files:       (various)
    currently installed:  28 Jul 2007
    latest available:     08 Oct 2007
Recommendation
    update ado-files
```

Stata compares the dates of your ado-files with those available from StataCorp. Here Stata is recommending that you update your official ado-files.

Important note: if you do not have write permission for C:\Program Files\Stata10, you cannot install official updates in this way. You may still download the official updates, but you will need to use the command-line version of `update`; see [U] **28 Using the Internet to keep up to date** for instructions.

Important note for Windows Vista users: you must start Stata in a special fashion to install official updates. Right-click on the Stata icon, and select **Run as Administrator** from the contextual menu. You must do this even if your account is an administrator's account. If you do not run Stata as an administrator, Vista will not allow you to install any of the files needed to update Stata.

Click on `update ado-files`, and Stata will update what is needed. For your information, Stata does not overwrite your original ado-files, which are installed in the directory C:\Program Files\Stata10\ado\base. Instead, Stata downloads updates to the directory C:\Program Files\Stata10\ado\updates. If there were ever a problem with the updates, you could simply remove the files in C:\Program Files\Stata10\ado\updates.

Updating the executable

The **Official Updates** system also checks your executable against the most current available:

---

update query

(contacting http://www.stata.com/)

```
Stata executable
    folder:               C:\Program Files\Stata10\
    name of file:         wsestata.exe
    currently installed:  28 Jul 2007
    latest available:     12 Nov 2007
```

Information about other types of updates may appear . . .

```
Recommendation
    update executable
```

Here Stata recommends that you update your executable.

Important note: if you do not have write permission for C:\Program Files\Stata10, you cannot install official updates in this way. You may still download the official updates, but you will need to use the command-line version of update; see [U] **28 Using the Internet to keep up to date** for instructions.

Important note for Windows Vista users: you must start Stata in a special fashion to install official updates. Right-click on the Stata icon, and select **Run as Administrator** from the contextual menu. You must do this even if your account is an administrator's account. If you do not run Stata as an administrator, Vista will not allow you to install any of the files needed to update Stata.

Click on update executable, and Stata will update what is needed. Stata copies the new executable to C:\Program Files\Stata10\wsestata.exe.bin; it does not replace your existing executable.

The final, crucial step is for you to install your new executable. Stata makes this easy for you. Exit all instances of Stata that are running except the one in which you ran update. Click the update swap link to complete the updating procedure. Stata will exit and start back up again.

Finding user-written programs by keyword

Stata has a built-in utility created specifically to search the Internet for user-written Stata programs. You can access it by selecting **Help > Search...**, choosing **Search net resources**, and entering a *keyword* in the field. Choosing **Help > SJ and User-written Programs** yields more specific choices for searching. The utility searches all user-written programs on the Internet, including the entire collection of *Stata Journal* and STB programs. The results are displayed in the Viewer, and you can click to go to any of the matches found.

For the syntax on how to use the equivalent search *keywords,* net command, see [R] **search**.

Downloading user-written programs

Downloading user-written programs is easy. Start by selecting **Help > SJ and User-written Programs**:

(Continued on next page)

```
╭─────────────────────────────────────────────────────────────────────────╮
│                                                                           │
│  Installation and maintenance of SJ, STB, and user-written programs       │
│  ──────────────────────────────────────────────────────────────────────  │
│                                                                           │
│       User-written programs -- SJ, STB, Statalist, and others -- are available │
│       from a variety of sources.  Use the links below to find, install, and │
│       uninstall them.                                                     │
│                                                                           │
│       New packages                                                        │
│                                                                           │
│           Search...            Search all sources.  Try this first!       │
│                                                                           │
│           Stata Journal        Programs from Stata Journal articles       │
│           STB                  Programs from Stata Technical Bulletin articles │
│           Statalist archive    Boston College Statalist program archive   │
│           Other locations      Other locations with available programs    │
│                                                                           │
│           Advanced                                                        │
│                   enter site name...                                      │
│                   cdrom drive                                             │
│                                                                           │
│       Previously installed packages                                       │
│                                                                           │
│           List                                                            │
│           Search...                                                       │
│           Update...                                                       │
│                                                                           │
│  Also see                                                                 │
│       Manual:  [U] 28 Using the Internet to keep up to date,              │
│                [R] net                                                    │
│       Online:  [R] adoupdate, [R] net, [R] search, [R] sj, stb, [R] update │
│                                                                           │
╰─────────────────────────────────────────────────────────────────────────╯
```

As the screen says: try Search... first.

Suppose that you were interested in finding more information or some user-written programs involving cubic splines. You select **Help > Search...**, select **Search all**, type cubic spline in the search box, and click the **OK** button.

```
╭─────────────────────────────────────────────────────────────────────────╮
│                                                                           │
│  . search cubic spline, all                                               │
│                                                                           │
│  Keyword search                                                           │
│                                                                           │
│           Keywords:  cubic spline                                         │
│             Search:  (1) Official help files, FAQs, Examples, SJs, and STBs │
│                      (2) Web resources from Stata and from other users    │
│                                                                           │
│  Search of official help files, FAQs, Examples, SJs, and STBs             │
│                                                                           │
│  [R]     mkspline . . . . . .  Linear and restricted cubic spline construction │
│          (help mkspline)                                                  │
│                                                                           │
│  [M-4]   mathematical  . . . . . . . . . . .  Important mathematical functions │
│          (help [M-4] mathematical)                                        │
│                                                                           │
│  [M-5]   spline3() . . . . . . . . . . . . . . . .  Cubic spline interpolation │
│          (help [M-5] spline3())                                           │
│                                                                           │
│  SJ-7-1  st0120  . Multivar. modeling with cubic reg. splines: A prin. approach │
│          (help mvrs, uvrs, splinegen if installed)  P. Royston and W. Sauerbrei │
│          Q1/07   SJ 7(1):45--70                                           │
│          discusses how to limit instability and provide sensible          │
│          regression models when using spline functions in a               │
│          multivariable setting                                            │
│                                                                           │
│  (etc.)                                                                   │
│                                                                           │
╰─────────────────────────────────────────────────────────────────────────╯
```

The first entry points to the built-in Stata command mkspline. You investigate this and find it interesting. You see the next two entries point to some built-in routines in Mata. You follow these links, because Mata is not only intriguing but also fast. You decide to check the fourth link, also. It points to an article in the *Stata Journal*, volume 7, number 1 (first quarter, 2007). You should click on the `st0120` link, as this will go to the programs associated with this article.

```
package st0120 from http://www.stata-journal.com/software/sj7-1

TITLE
      SJ7-1 st0120.  Multivariable regression spline models
DESCRIPTION/AUTHOR(S)
      Multivariable regression spline models
      by Patrick Royston, UK Medical Research Council
         Willi Sauerbrei, University Medical Center, Freiberg, Germany
      Support:  patrick.royston@ctu.mrc.ac.uk
      After installation, type help mvrs, uvrs, and splinegen
INSTALLATION FILES                              (click here to install)
      st0120/mvrs.ado
      st0120/mvrs.hlp
      st0120/splinegen.ado
      st0120/splinegen.hlp
      st0120/uvrs.ado
      st0120/uvrs.hlp
ANCILLARY FILES                                 (click here to get)
      st0120/brca_curves.do
      st0120/brca_gof2.do
      st0120/brca_gof.do
      st0120/ex1.do
      st0120/ex2.do
      st0120/fig_mcyc.do
      st0120/g.do
      st0120/brcancer.dta
      st0120/mcyc.dta
```

There are many files here! You can click on the help files to see if the commands here look like they are interesting. If you decide that you would like to try the commands, you can click the link `click here to install`. If you decide that you would like to use some of the ancillary files—files that typically help explain the workings of the command, you could install those, too. You do not need to worry—doing so will not interfere in any way with your copy of Stata. We will show you how to uninstall these programs safely shortly.

Clicking on the install link yields the following output.

```
checking st0120 consistency and verifying not already installed...
installing into c:\ado\plus\...
installation complete.
```

That's all there is to installing a user-written package.

(Continued on next page)

Now suppose that you decide that you would like to uninstall the package. Doing so is simple enough: select **Help > SJ and User-written programs**, and click the List link. You should see the following:

```
[1] package st0120 from http://www.stata-journal.com/software/sj7-1
      SJ7-1 st0120.  Multivariable regression spline models
```

If you click on the one-line description of the program, you will see the full description of what has been installed.

```
[1] package st0120 from http://www.stata-journal.com/software/sj7-1

TITLE                                     (click here to uninstall)
      SJ7-1 st0120.  Multivariable regression spline models
DESCRIPTION/AUTHOR(S)
      Multivariable regression spline models
      by Patrick Royston, UK Medical Research Council
         Willi Sauerbrei, University Medical Center, Freiberg, Germany
      Support:  patrick.royston@ctu.mrc.ac.uk
      After installation, type help mvrs, uvrs, and splinegen
INSTALLATION FILES
      m/mvrs.ado
      m/mvrs.hlp
      s/splinegen.ado
      s/splinegen.hlp
      u/uvrs.ado
      u/uvrs.hlp
INSTALLED ON
      15 Apr 2007
```

You can uninstall materials by clicking on click here to uninstall when you are looking at the package description.

Try it.

For information on downloading user-written programs by using the net command, see [R] **net**.

A   Troubleshooting Stata

Contents

A.1   If Stata does not start

You tried to start Stata and it refused; Stata or your operating system presented a message explaining that something is wrong. Here are the possibilities:

Cannot find license file

This message means just what it says; nothing is too seriously wrong. Stata simply could not find what it is looking for. The two most common reasons for this are that you did not complete the installation process or Stata is not installed where it should be.

Did you insert the codes printed on your paper license to unlock Stata? If not, go back and complete the initialization procedure.

Assuming that you did unlock Stata, Stata is merely mislocated, or the location has not been filled in.

Error opening or reading the file

Something is distinctly wrong and for purely technical reasons. Stata found the file that it was looking for, but either the operating system refused to let Stata open it or there was an I/O error. About the only way this could happen would be a hard-disk error. Stata technical support will be able to help you diagnose the problem; see [U] **3.9 Technical support**.

License not applicable

Stata has determined that you have a valid Stata license, but the license does not apply to the version of Stata that you are trying to run. You would get this message if, for example, you tried to run Stata for Windows with a Stata for Macintosh license.

The most common reason for this message is that you have a license for Stata/IC but you are trying to run Stata/SE, or you have a license for Small Stata but you are trying to run Stata/IC or Stata/SE. If this is the case, reinstall Stata, making sure to choose the appropriate flavor.

Other messages

The other messages indicate that Stata thinks you are attempting to do something that you are not licensed to do. Most commonly, you are attempting to run Stata over a network when you do not have a network license, but there are many other alternatives. There are two possibilities: either you really are attempting to do something that you are not licensed to do, or Stata is wrong. In either case, you are going to have to contact us. Your license can be upgraded, or, if Stata is wrong, we can provide codes to make Stata stop thinking that you are violating the license; see [U] **3.9 Technical support**.

A.2   Troubleshooting tips

Crashes are called *application faults* in Windows. The dreaded application fault is not always the application's fault. It can be caused by configuration problems, bugs in device drivers, memory conflicts, and even hardware problems.

If you experience an application fault, first make sure that you are running the most current version of Stata (see chap. 20 for information on updating). If the problem still exists, look at the frequently asked questions (FAQ) for Windows in the user-support section of the Stata web site, http://www.stata.com/. You may find the answer to the problem there. If not, we can help, but you must give us as much information as possible.

Reboot your computer, restart Stata, and try to reproduce the fault, writing down everything you do before the fault occurs. We will want that information, along with the contents of your CONFIG.SYS, AUTOEXEC.BAT, SYSTEM.INI, and WIN.INI files. If you cannot email them to us, at least print them so that you can look at them when you call us.

If Stata used to work on your computer but suddenly stopped working, try to remember any hardware or software that you have recently installed.

Also, give us as much information about your computer as possible. What version of Windows are you running? How much memory do you have? What processor do you have? What brand is your computer? What kind of mouse do you have? What kind of video card? Finally, we need your Stata serial number and the date your version of Stata was "born". Include them if you email, and know them if you call. You can obtain them by typing about in Stata's Command window. about lets us know everything about your copy of Stata, including the version and the date it was produced.

B Managing memory

Contents

B.1 Memory size considerations

Stata works with a copy of the dataset that it loads into memory.

By default, Small Stata allocates about 300 KB to Stata's data areas, and you cannot change it.

By default, Stata/IC allocates 1 MB to Stata's data areas, and you can change it.

By default, Stata/MP and Stata/SE allocate 10 MB to Stata's data areas, and you can change it.

You can even change the allocation to be larger than the physical amount of memory on your computer because Windows provides virtual memory.

Virtual memory is slow but adequate in rare cases when you have a dataset that is too large to load into real memory. If you use large datasets often, we recommend that you add more memory to your computer. See [GSW] **B.4 Virtual memory and speed considerations**.

You can change the allocation when you start Stata, as will be discussed in [GSW] **C.4 Specifying the amount of memory allocated**.

You can also change the total amount of memory allocated while Stata is running and optionally make that setting the default to be used by future invocations of Stata. That is the topic of this chapter.

It does not much matter which method you use. Being able to change the total on the fly is convenient, but even if you cannot do this, it just means that you must specify it ahead of time, and if later you need more, you must exit Stata and reinvoke it with the larger total.

Do not specify more memory than you actually need. Doing so does not benefit Stata; in fact, it may slow Stata down!

B.2 Setting the size on the fly

Assume that you have changed nothing about how Stata starts, so you have the default amount of memory (10 MB for Stata/MP and Stata/SE, 1 MB for Stata/IC) allocated to Stata's data areas. You are working with a large dataset and now wish to increase memory to 32 MB. You can type

```
. set memory 32m
```

and, if your operating system can provide the memory to Stata, Stata will work with the new total. Later in the session, if you want to release that memory and work with only 2 MB, you could type

```
. set memory 2m
```

133

There is only one restriction on the `set memory` command: whenever you change the total, there cannot be any data already in memory. If you have a dataset in memory, you save it, clear memory, reset the total, and then use it again. We are getting ahead of ourselves, but you might type

```
. save mydata, replace
file mydata.dta saved

. clear

. set memory 32m
...

. use mydata
```

When you request the new allocation, your operating system might refuse to provide it:

```
. set memory 512m
op. sys. refuses to provide memory
r(909);
```

If that happens, you are going to have to take the matter up with your operating system. In the above example, Stata asked for 512 MB, and the operating system said no.

B.3   The memory command

`memory` helps you figure out whether you have enough memory to do something. If you are using Stata/SE (or Stata/MP), you could type `memory`:

```
. memory
```

	bytes	
Details of **set memory** usage		
overhead (pointers)	114,136	10.88%
data	913,088	87.08%
data + overhead	1,027,224	97.96%
free	21,344	2.04%
Total allocated	1,048,568	100.00%
Other memory usage		
set maxvar usage	1,816,666	
set matsize usage	1,315,200	
programs, saved results, etc.	772	
Total	3,132,638	
Grand total	4,181,206	

You can type `memory` in Stata/IC, too, but the output will vary slightly from that of Stata/SE and Stata/MP; see [D] **memory**. Having 21,344 bytes free is not much. You might increase the amount of memory allocated to Stata's data areas by specifying `set memory 2m`.

```
. save nlswork
. clear
. set memory 2m
...

. use nlswork
(NLS Women 14-26 in 1968)
```

```
. memory
```

	bytes	
Details of **set memory** usage		
overhead (pointers)	114,136	5.44%
data	913,088	43.54%
data + overhead	1,027,224	48.98%
free	1,069,920	51.02%
Total allocated	2,097,144	100.00%
Other memory usage		
set maxvar usage	1,816,666	
set matsize usage	1,315,200	
programs, saved results, etc.	773	
Total	3,132,639	
Grand total	5,229,783	

More than 1 MB is now free; that's better. See [D] **memory** for more information.

You can use `set memory` to set the memory for future Stata sessions as well as the current session by using the `permanently` option. If, for example, you wanted 32 MB of memory allocated for data in future Stata sessions, you could type

```
. set memory 32m, permanently
```

B.4 Virtual memory and speed considerations

When you open (`use`) a dataset in Stata for Windows, Stata loads the entire dataset into memory.

When you use more memory than is physically available on your computer, Stata slows down. If you are using only a little more memory than is on your computer, performance is probably not too bad. On the other hand, when you are using a lot more memory than is on your computer, performance will be noticeably affected. We recommend then that you

```
. set virtual on
```

Virtual memory systems exploit locality of reference, which means that keeping objects closer together allows virtual memory systems to run faster. `set virtual` controls whether Stata should perform extra work to arrange its memory to keep objects close together. By default, `virtual` is set `off`. `set virtual` can be specified only if you are using Stata/MP, Stata/SE, or Stata/IC for Windows.

In general, you want to leave `set virtual` set to the default of `off` so that Stata will run faster.

When you `set virtual on`, you are asking Stata to arrange its memory so that objects are kept closer together. This setting requires Stata to do a substantial amount of work. We recommend setting virtual on only when the amount of memory in use drastically exceeds what is physically available. In these cases, setting virtual on will help, but keep in mind that performance will still be slow. See [U] **6.5 Virtual memory and speed considerations**. If you are using virtual memory often, you should consider adding memory to your computer.

Notes

C   Advanced Stata usage

Contents

C.1   The Windows Properties Sheet

When you double-click on a shortcut to start an application in Windows, you are actually executing instructions defined in the shortcut's Properties Sheet. To open the Properties Sheet for any shortcut, right-click on the shortcut, and select **Properties**.

Open the Properties Sheet for Stata's shortcut. Click on the **Shortcut** tab. You will see something that contains the following information:

Stata/SE

Target type:	Application
Target location:	Stata 10
Target:	`C:\Program Files\Stata10\wsestata.exe`
Start in:	`C:\data\`
Shortcut key:	`None`
Run:	`Normal window`

The field names may be slightly different, depending on the version of Windows that you are running. The names and locations of files may vary from this. There are two things to pay attention to: the **Target** and **Start in**. **Target** is the actual command that is executed to invoke Stata. **Start in** is the directory to switch to before invoking the application. You can change these fields and then click **OK** to save the updated Properties Sheet.

You can have Stata start in any directory you desire. Just change the **Start in** field of Stata's Properties Sheet. Of course, once Stata is running, you can change directories whenever you wish; see [D] **cd**.

C.2   Making shortcuts

You can arrange to start Stata without going through the **Start** menu by creating a shortcut on the Desktop. The easiest way to do this is to copy the existing Stata shortcut to the Desktop. You can also create a shortcut directly from the Stata executable. Here are the details:

1. Open the `C:\Program Files\Stata10` folder or the folder where you installed Stata.

2. In the folder, find the executable for which you want a new shortcut. The filenames are

Stata/MP:	`wmpstata.exe`
Stata/SE:	`wsestata.exe`
Stata/IC:	`wstata.exe`
Small Stata:	`wsmstata.exe`

Right-click and drag the appropriate executable onto the Desktop.

3. Release the mouse button, and select **Create Shortcut(s) Here** from the menu that appears.

You have now created a shortcut. If you want the shortcut in a folder rather than on the Desktop, you can drag it into whatever folder appeals to you.

You set the properties for this shortcut just as you would normally. Right-click on the shortcut, and select **Properties**. Edit the Properties Sheet as explained in [GSW] **C.1 The Windows Properties Sheet**. Some users create separate shortcuts for starting Stata with different amounts of memory.

C.3 Executing commands every time Stata is started

Stata looks for the file `profile.do` when it is invoked and, if it finds it, executes the commands in it. Stata looks for `profile.do` first in the directory where Stata is installed, then in the current directory, then along your PATH, then in your home directory as defined by Windows' USERPROFILE environment variable, and finally along the ado-path (see [P] **sysdir**); we recommend that you put `profile.do` in default working directory, `C:\data`.

Say that every time you start Stata you want `matsize` set to 100 (see [R] **matsize**). Create the file `profile.do` in `C:\data` containing

```
set matsize 100
```

When you invoke Stata, this command will be executed:

```
(usual opening appears, but with the addition)
running C:\data\profile.do ...

. _
```

```
(usual opening appears, but with the addition)
running /home/mydir/bin/profile.do ...

. _
```

You could also type `set matsize 100, permanently` in Stata. The `permanently` option tells Stata to remember the setting for future sessions and eliminates the need to put the `set matsize` command in `profile.do`.

`profile.do` is treated just as any other do-file once it is executed; results are just the same as if you started Stata and then typed **run profile.do**. The only special thing about `profile.do` is that Stata looks for it and runs it automatically.

System administrators may also find `sysprofile.do` useful. This file is handled in the same way as `profile.do`, except that Stata looks for `sysprofile.do` first. If that file is found, Stata will execute any commands it contains. After that, Stata will look for `profile.do` and, if that file is found, execute the commands in it.

One example of how `sysprofile.do` might be useful is for system administrators who want to change the path to one of Stata's system directories. Here `sysprofile.do` could be created to contain the command

```
sysdir set SITE "\\Matador\StataFiles"
```

See [U] **16 Do-files** for an explanation of do-files. They are nothing more than ASCII text files containing a sequence of commands for Stata to execute.

C.4   Specifying the amount of memory allocated

Only users of Stata/MP, Stata/SE, and Stata/IC may change the amount of memory Stata allocates.

You can change the amount of memory once Stata is running by using the `set memory` command, and you can make that setting permanent; see [GSW] **B. Managing memory**.

C.5   Launching Stata by double-clicking a Stata file

The first time that you start Stata for Windows, Stata registers with Windows the actions to perform when you double-click on certain types of files. You can then start Stata by double-clicking on a Stata `.dta` dataset, a Stata `.do` do-file, or a Stata `.gph` graph file. In all cases, your working directory will become the folder containing the file that you have double-clicked.

Stata will behave as you would expect in each case. If you double-click a dataset, Stata will open the dataset after it starts. If you double-click a do-file, the do-file will be run by Stata. If you double-click a graph, the graph will be opened by Stata.

When you right-click on a Stata `.gph` file, you can select **Print** or **Print to default**. If you select **Print**, Stata launches, displays the graph, and opens a print dialog. If you select **Print to default**, Stata launches and prints the graph to the default printer, bypassing the print dialog. When the graph is finished printing (or when you cancel the print), Stata exits.

C.6   Stata batch mode

You can run large jobs in Stata in batch mode. To do so, open a Command Prompt window, change to your data directory, and type

```
C:\data> "C:\Program Files\Stata10\wsestata.exe" /b do bigjob
```

This tells Stata to execute the commands in `bigjob.do`, suppress all screen output, and route the output to `bigjob.log` in the same directory. If you desire a SMCL log file rather than an ASCII file, specify `/s` rather than `/b`.

If the do-file loads datasets that require more than the default amount of memory (10 MB for Stata/MP and Stata/SE, 1 MB for Stata/IC), you will need to allocate this to Stata when you start it. Typing

```
C:\data> "C:\Program Files\Stata10\wsestata.exe" /m15 /b do bigjob
```

will run `bigjob.do` with 15 MB of memory. `bigjob.do` can also try to increase the memory allocated to Stata while it is running with the `set memory` command; see [GSW] **B. Managing memory**.

While the do-file is executing, the Stata icon will appear on the taskbar together with a rough percentage of how much of `bigjob.do` Stata has executed. (Stata calculates this percentage from the number of characters in `bigjob.do`, so the percentage may not accurately reflect the amount of time left for the job to complete.)

If you click the icon on the taskbar, Stata will display a box asking if you want to cancel the batch job.

Once the do-file is complete, Stata will flash the icon on the taskbar on and off. You can then click the icon to close Stata. If you wish for Stata to automatically exit after running the batch do-file, use /e rather than /b.

You do not have to run large do-files in batch mode. Any do-file that you run in batch mode can also be run interactively. Simply start Stata, type log using *filename*, and type do *filename*. You can then watch the do-file run, or you can minimize Stata while the do-file is running.

C.7   Running simultaneous Stata sessions

Each time you double-click on the Stata icon or launch Stata in any other way, you invoke a new instance of Stata, so if you want to run multiple Stata sessions simultaneously, you may. The title bar of each new Stata that is invoked will reflect its instance number.

D More on Stata for Windows

D.1   Using Stata datasets and graphs created on other platforms

Stata will open any Stata `.dta` dataset or `.gph` graph file regardless of the platform on which it was created, even if it was a Macintosh or Unix system. Also, Stata for Macintosh and Stata for Unix users can use any files that you create. If you transfer a Stata file by using file transfer protocol (FTP), just remember to transfer using binary mode rather than ASCII.

D.2   Exporting a Stata graph to another document

Suppose that you wish to export a Stata graph to a document created by your favorite word processor or desktop publishing (DTP) application. You have two main choices for exporting graphs: you may copy and paste the graph by using the Clipboard, or you may save the graph in one of several formats and import the graph into the application.

D.2.1   Exporting the graph by using the Clipboard

The easiest way to export a Stata graph into another application is via the Clipboard.

Either create your graph or redisplay an existing graph. To copy it to the Clipboard, right-click on the Graph window, and select **Copy**. Stata will copy the graph as an Enhanced Metafile (EMF); this ensures that the receiving application obtains it in the highest resolution possible. If the receiving application does not understand the EMF format, select **Edit > Preferences > Graph Preferences**, click on the **Clipboard** tab, and click the *Windows Metafile (WMF)* radio button to copy graphs in the Windows Metafile (WMF) format before copying.

A metafile contains the commands necessary to redraw the graph. That is, a metafile is a collection of lines, points, text, and color information. Metafiles, therefore, can be edited in a structured drawing program.

After you have copied the graph to the Clipboard, switch to the application into which you wish to import the graph and paste it. In most applications, this is accomplished by selecting **Edit > Paste**. Consult the documentation for your particular application for more details.

(Continued on next page)

D.2.2   Exporting the graph to a file

Stata can export graphs to several different file formats. If you right-click on a graph, select **Save Graph...**, and then click the **Save as type** combobox, you will see that Stata can save in Windows Metafile, Enhanced Metafile, Portable Network Graphics (PNG), TIFF, PostScript, Enhanced PostScript (EPS), and EPS with TIFF preview. Windows Metafile, Enhanced Metafile, PostScript, and EPS formats are vector formats, whereas PNG and TIFF are bitmap formats. EPS files are recommended when you want to export your graph to an application on another platform, such as TeX or LaTeX on Unix (which is how all the graphs in Stata's manuals were created), or for best output on a PostScript printer. If you wish to include a preview of the graph so that it may be viewed in your DTP application, choose **EPS with TIFF preview**. Choosing the preview option does not affect how the graph is printed. PNG is well suited for placing graphs on a web page. See the *Graphics Reference Manual* for more information.

D.3   Installing Stata for Windows on a network drive

You will need a network license before you can install Stata on a network drive. You can install Stata from the server, or if you have the appropriate privileges, you can install Stata directly to the network drive.

Follow the instructions for installing Stata in chapter 1. Depending on where you wish to provide network access to Stata from, you may have to change the installation directory at step 5 of the installation instructions.

Once Stata is installed, run it to initialize the license. Mount the network drive that Stata is installed on from a workstation. Right-click on the Desktop or on the Windows **Start** menu, and select **New > Shortcut**. Type the path for the Stata executable into the edit field, or click **Browse...** to locate it. Enter Stata for the name of the shortcut.

Once a shortcut for Stata has been created, right-click on it, and select **Properties**. Set the default working directory for Stata by changing the **Start in:** field to a local drive that users have write access to. This is where Stata will store datasets, graphs, and other Stata-related files. If the workstation will be used by more than one user, consider changing the **Start in:** edit field to the environment variable `%HOMEDRIVE%%HOMEPATH%`. Doing so will set the default working directory to each user's home directory.

D.4   Changing a Stata for Windows license

If you've already installed Stata and your license needs to be changed,

1. You will need the *License and Authorization Key*.

2. Open the Stata directory from the hard drive.

3. Right-click on the `STATA.LIC` file, and select **Rename**.

4. Change the name of the file to `STATA.LIC.old`.

5. Run Stata, which will then prompt you for your license information and recreate the `STATA.LIC` file.

6. Run Stata as an unprivileged user to ensure that it runs correctly. If Stata runs for the privileged user but does not run for the unprivileged user, check the permissions on the `STATA.LIC` file.

7. Once Stata runs properly with the new license information, delete the file `STATA.LIC.old`.